Neuroimmunoendocrinology

Chemical Immunology

Vol. 52

KARGER

Basel · München · Paris · London · New York · New Delhi · Bangkok · Singapore · Tokyo · Sydney

Neuroimmunoendocrinology

2nd, revised and enlarged edition

Volume Editor
J. Edwin Blalock, Birmingham, Ala.

22 figures, 3 color plates and 14 tables, 1992

KARGER

Basel · München · Paris · London · NewYork · New Delhi · Bangkok · Singapore · Tokyo · Sydney

Chemical Immunology

Formerly published as 'Progress in Allergy'
Founded 1939 by Paul Kallós

Vol. 43 (1st edition)
Neuroimmunoendocrinology
Editors: J. Edwin Blalock; Kenneth L. Bost, Birmingham, Ala.
X+166 p., 17 fig., 1 cpl., 13 tab., hard cover, 1988. ISBN 3-8055-4774-9

Bibliographic Indices
This publication is listed in bibliographic services, including Current Contents® and Index Medicus.

Drug Dosage
The authors and the publisher have exerted every effort to ensure that drug selection and dosage set forth in this text are in accord with current recommendations and practice at the time of publication. However, in view of ongoing research, changes in government regulations, and the constant flow of information relating to drug therapy and drug reactions, the reader is urged to check the package insert for each drug for any change in indications and dosage and for added warnings and precautions. This is particularly important when the recommended agent is a new and/or infrequently employed drug.

Printed on acid-free paper.
ISBN 3-8055-5488-5

Contents

Introduction to the 1st Edition

The study of interactions between the immune and neuroendocrine systems is a currently popular and rapidly advancing field which had its foundation in anecdotal observations of the association between 'personality' and disease. A measure of scientific credence was afforded to the area by the observation that, like many other physiologic responses, immune reactions could be conditioned in a classical Pavlovian fashion [1]. A possible mechanism was then found in Selye's [2] observation of thymic involution during 'stress'. The concept of the effects of stress on immunity turned out to be a two-edged sword. On the one hand, it provided a molecular basis, in the form of adrenal glucocorticoids, for neuroendocrine control of immunity. On the other hand, it led to the general and often currently held notion that steroid hormones are the sole players in neuroendocrine modulation of the immune system. Of course, recent studies have shown that this is clearly not the case since stressed adrenalectomized animals are functionally immunosuppressed [3]. Another impediment to the development of the area was the ability to have immunologic responses proceed in vitro. This inadvertently led to the idea that the immune system is a totally autonomous and self-regulating unit. If this were the case, then the immune system would be unlike all other organ systems. Furthermore, this view overlooks the rich hormonal milieu in which many in vitro immune responses occur. The end result of this series of events is the thought that if immune neuroendocrine interactions occur, they are mediated by steroid hormones. The picture has been further clouded by the predominance of studies of the psychological aspects of immune neuroendocrine interactions. Though it is not necessarily our view, such studies have given the field an aura of being a 'soft' science, thus the intent of *Neuroimmunoendocrinology* is to highlight the cellular and molecular aspects of this field.

References

1. Ader R: A historical account of conditioned immunobiologic responses; in Ader (ed): Psychoneuroimmunology. New York, Academic Press, 1981, pp 321–349.
2. Selye H: Thymus and adrenals in the response of the organism to injuries and intoxications. Br J Exp Pathol 1936;17:234–248.
3. Keller SE, Weiss JM, Schleifer SJ, Miller NE, Stein M: Stress-induced suppression of immunity in adrenalectomized rats. Science 1983;221:1301–1304.

J. Edwin Blalock, PhD, Department of Physiology and Biophysics,
University of Alabama at Birmingham, Birmingham, AL 35294-0005 (USA)

Introduction to the 2nd Edition

In the 3 years since the first publication of this volume, we have witnessed an explosion of information on immune neuroendocrine interactions. This has been evidenced by numerous international congresses, the inclusion of many symposia at the annual meetings of immunology, endocrinology, and neuroscience societies and the initiation of at least two new journals on the subject (i.e., *Progress in NeuroEndocrinImmunology* and *Advances in Neuroimmunology*). Among the highlights which have fueled the scientific growth of this discipline are: the tremendous increase in the number of neuroendocrine hormones and peptide neurotransmitters as well as their receptors which are endogenous to the immune system; the finding of cytokines such as IL-1 and IL-6 as well as their receptors in neural and endocrine tissues; and the profound effects of bidirectional communication between the immune and neuroendocrine systems on an animal's physiology. As an example, it is interesting to note that IL-1 is probably a more potent activator of the hypothalamic pituitary adrenal axis than is corticotropin releasing factor (CRF). Thus the paradox that without knowing it immunologists discovered a most potent 'hypothalamic releasing factor'. Contrariwise, CRF is a very effective inducer of IL-1. Thus, endocrinologists without knowing it discovered a cytokine. Such are the strange but exciting ironies of the development of a new discipline.

As a result of the great accumulation of new knowledge and the overwhelming success of the first edition, myself, and the coauthors have written the present revised edition of *Neuroimmunoendocrinology*. Once again, the emphasis is on the molecular, cellular, and physiologic aspects of immune neuroendocrine interactions. Because of the expansion of the literature in this area, most chapters and bibliographies are longer. It is our fondest hope that this volume will be a valuable resource to the evolution of this exciting new discipline.

J. Edwin Blalock, PhD, Department of Physiology and Biophysics,
University of Alabama at Birmingham, Birmingham, AL 35294–0005 (USA)

Blalock JE (ed): Neuroimmunoendocrinology, 2nd rev ed.
Chem Immunol. Basel, Karger, 1992, vol 52, pp 1–24

Production of Peptide Hormones and Neurotransmitters by the Immune System

J. Edwin Blalock

Department of Physiology and Biophysics,
University of Alabama at Birmingham, Ala., USA

Introduction

Among the newly recognized similarities between the immune and neuroendocrine systems is the observation in both systems of an increasing number of peptides and proteins that are used as intercellular messengers. Previous to the work which is reviewed herein, these molecules were largely deemed to be endogenous informational substances for the system from which they were originally isolated. Recent evidence, however, strongly supports the notion that neuroendocrine peptides are endogenous to the immune system and are used for both intraimmune system regulation, as well as for bidirectional communication between the immune and neuroendocrine systems. In fact, in the 3 years since the first edition of this volume, the number of neuroendocrine peptide hormones and peptide neurotransmitters found in the immune system has more than doubled (table 1).

Production and Structure of Leukocyte-Derived Peptide Hormones

Pituitary Hormones

Ironically, corticotropin (ACTH) was one of the first neuroendocrine peptide hormones to be described and was the first de novo synthesized hormone to be found in the immune system [1, 2]. ACTH is a 39 amino acid peptide whose production is primarily associated with the pituitary gland and whose classical action is to elicit a glucocorticoid response from the adrenal glands during times of stress. In the immune system, human peripheral blood lymphocytes and mouse spleen cells were initially observed to

Table 1. Stimuli for immunologically-derived peptide hormones and neurotransmitters

Hormone or neurotransmitter	Stimuli	Reference
Corticotropin (ACTH)	Newcastle disease virus	1,2,8,9
	Lymphotropic viruses	11
	Bacterial lipopolysaccharide	3,13,80,88,89
	Corticotropin releasing factor (CRF)	47,70,71
	Arginine vasopressin	47
	Interleukin-1	71
	Thymopentin	74
	Unknown	5
Endorphins	Newcastle disease virus	1,2,9
	Bacterial lipopolysaccharide	3,14,79,95
	Corticotropin releasing factor	47,70,71
	Arginine vasopressin	47
	Interleukin-1	71
	Psychoactive drugs	69
	Cold water stress	96
Thyrotropin (TSH)	Staphylococcal enterotoxin A	17
	Thyrotropin releasing hormone (TRH)	18,19
Chorionic gonadotropin (CG)	Mixed lymphocyte reaction	27
Luteinizing hormone (LH)	Luteinizing hormone releasing hormone	28–31,34
Follicle-stimulating hormone (FSH)	Concanavalin A	35,36
Growth hormone (GH)	Unknown	37
	Growth hormone releasing hormone (GHRH)	38
	Growth hormone	39
Prolactin	Concanavalin A	20–22
	Unknown	23–25
Corticotropin releasing factor	Unknown	42
Growth hormone releasing hormone	Unknown	41,46
Luteinizing hormone releasing hormone	Unknown	43–45
[Met]enkephalin	Concanavalin A	16
Arginine vasopressin	Unknown	56,57

Table 1. (Continued)

Hormone or neurotransmitter	Stimuli	Reference
Oxytocin	Unknown	56,58
Neuropeptide Y	Unknown	54
Vasoactive intestinal peptide	Histamine liberators	48
	Hypersensitivity	51
	Unknown	52,53
Somatostatin	Hypersensitivity	51
	Unknown	50,52
Substance P	Unknown	55
Parathyroid hormone-related protein	Human T cell lymphotropic virus I	60
Calcitonin gene-related peptide	Hypersensitivity	51
Insulin-like growth factor	Growth hormone	63
Suppressin	Concanavalin A	66
Neuroimmune protein	Developmental	68
	Concanavalin A	67
	Bacterial lipopolysaccharide	
Neuroimmune protein	Bacterial lipopolysaccharide	67

express an ACTH-like peptide following virus infection or interaction with transformed cells or bacterial lipopolysaccharide (LPS) [1–3]. In contrast to the inducible synthesis of ACTH and endorphins by most lymphocytes, a subpopulation of mouse splenic macrophages, as well as some rat lymphocytes in the tunica propria, produced these peptides in a constitutive fashion [4, 5]. Pituitary ACTH is derived from a large precursor protein, proopiomelanocortin (POMC), which contains the endogenous opioid peptides, endorphins [6]. Likewise, leukocyte-derived endorphins are coordinately expressed along with ACTH, and this is most likely due to their derivation from POMC. In fact, there would seem to be no other source for these peptides, since all ACTH and endorphins are derived from a single POMC gene. Among the many shared characteristics between immune system and pituitary-derived ACTH and endorphins are: shared antigenicity as determined with monospecific antibodies against synthetic peptides hormones;

identical retention times on reverse-phase, high-pressure liquid chromatographic (HPLC) columns; identical molecular weights; and shared biological activities [7, 8] and the presence of POMC-related mRNA in lymphocytes and macrophages [5, 9–15]. Most recently, identity between pituitary and leukocyte ACTH has been unequivocally established by demonstrating that the amino acid and nucleotide sequences of mouse splenic and pituitary ACTH were identical [13, 14]. Thus, there is no longer any doubt that cells of the immune system make authentic POMC-derived peptides. Cells of the immune system also seem to synthesize another family of endogenous opiates, the enkephalins, that are related to the endorphins. Abundant levels of preproenkephalin mRNA have been found in mitogen or antigen-activated but not resting T helper (h) cells. This mRNA represented from 0.1 to 0.5% of the total mRNA depending on the particular T_h line which was induced. The mRNA was apparently translated and the product was secreted since immunoreactive (ir) [Met] enkephalin was detected in T_h cell culture supernates [16].

Thyrotropin (TSH) was the first de novo synthesized glycoprotein hormone to be found in the immune system. Its induction was initially observed in response to activation of human peripheral blood cells by a particular T cell mitogen, staphylococcal enterotoxin A (SEA) [17]. More recently, TSH has been shown to be constitutively produced by human T cell leukemia lines [18], In these studies, the lymphocyte-derived TSH was recognized by a monospecific antibody to TSH-β and was shown to have the same molecular weight as pituitary TSH. Further, the intact molecule was shown to be composed of two polypeptide chains of the molecular weight of TSH-α and TSH-β. Whether the immune system-derived hormone is biologically active and will stimulate thyroid cells remains to be established. With regard to identity with the pituitary hormone, TSH-β-related mRNA has been observed in human and mouse lymphocytes [18, 19].

Yet another T cell mitogen, concanavalin A (ConA), was reported to cause lymphocyte production of a prolactin-related mRNA [20] and prolactin-like molecules [21]. The mRNA (10 kb) was larger than that in the pituitary gland [20] and the lymphocyte associated prolactin-like molecule had a molecular weight of 48 kDa compared to pituitary prolactin of 23–24 kDa [22]. Friesen and co-workers [23–25] have conclusively demonstrated the presence of prolactin and its mRNA in an IM9 human B lymphoblastoid cell line. The IM9 mRNA was 150 nucleotides longer than the pituitary transcript due to an elongation of the 5′ untranslated region. This extension resulted from the use of a new 5′ noncoding exon that was 5–7 kb upstream of the human pituitary prolactin gene exon 1. In contrast to de novo synthesis, Clevenger et al. [26] recently reported that IL-2 stimulated T lymphocytes accumulated prolactin by internalization from the culture medium. Thus at present it would appear that there are at least three mechanisms for prolactin association with cells of the immune system. These are: (1) active uptake from extracellular fluids; (2) constitutive production by lymphoid cells, and (3) antigen- or mitogen-induced synthesis from lymphocytes.

Interestingly, mixed lymphocyte reactions (MLR) result in T cell mitogenesis but unlike SEA or ConA do not evoke the production of TSH or prolactin. Rather, this allogeneic stimulus results in the production of an ir chorionic gonadotropin (CG) [27]. CG production, as monitored by immunofluorescence with antibody to CGβ, paralleled the blastogenic response of the MLR. Gel filtration of the de novo synthesized lymphocyte-derived CG showed that this material comigrated with the human (h)CG standard at a molecular weight of approximately 58 kDa. This molecule was apparently glycosylated since it bound to a ConA affinity column. The lymphocyte-derived CG was also dissociable into two subunits of the molecular weight of CGα and β, 32 and 18 kDa respectively. The material from the MLR was biologically active since it elicited testosterone production from Leydig cells and this activity was neutralized by antiserum to CG. The finding that mouse, as well as human, lymphocytes produce CG is important since controversy lingers as to whether mouse placentas produce the molecule. The results with the murine MLR would seem to support the notion that rodents have the equivalent of hCG.

Besides CG, two other gonadotropins have recently been found in the immune system. In response to luteinizing hormone (LH) releasing hormone (RH), human [28, 29], mouse [30. 31] and porcine [32–34] lymphocytes were shown to synthesize and release irLH. The lymphocyte-derived LH shared antigenicity, molecular mass, subunit structure, glycosylation and reverse-phase HPLC profile with the pituitary hormone [28]. Most recently, the amino terminal sequence of the lymphocyte LH β chain was shown to be identical to LH β [E.M. Smith, pers. commun.]. An ir follicle-stimulating hormone (FSH) was also detected in cultured rat lymphocytes [35]. The levels of this hormone were markedly elevated by a T cell mitogen (ConA) and the biologically active irFSH existed in two molecular weight forms. One form was quite similar to pituitary FSH (30 kDa) while the other (54 kDa) was not [36]. Collectively, these data show that a number of different animal species leukocytes can produce the three principal gonadotropins (i.e. CG, LH, and FSH).

All of the aforementioned peptide hormones have required an inductive event for their synthesis by lymphocytes. In contrast, splenic lymphocytes spontaneously produce growth hormone (GH) mRNA after about 4–6 h of in vitro culture [37]. Thus in vivo the lymphocyte GH gene must be under tonic suppression. In spite of this spontaneous production, GHRH [38] as well as GH [39] marginally augmented lymphocyte GH synthesis. GH-related RNA has been detected in cells from spleen, thymus, bone marrow, and peripheral blood [40]. The mRNA has the same molecular weight as GH mRNA and is translated into a 22-kDa biologically active irGH [37]. In all likelihood, the primary sequence of the immunocyte-derived GH will be identical to that of the pituitary since GH-related cDNA from lymphocytes gives an identical restriction endonuclease map to that of pituitary GH cDNA [41].

Hypothalamic Releasing Factors

In general, each pituitary hormone is positively regulated by a given releasing hormone from the hypothalamus. Evidence has recently accumulated suggesting that a similar situation may exist in the immune system. Stephanou et al. [42] were the first to demonstrate a hypothalamic releasing factor-like peptide in leukocytes. They found an ir corticotropin releasing factor (CRF) in human lymphocytes and neutrophils which eluted earlier than standard CRF on HPLC. The lymphocyte CRF-like mRNA was 1.7 kb in comparison with the 1.5 kb mRNA for hypothalamic CRF. In contrast to the irCRF, Emanuele et al. [43] demonstrated a biologically active irLHRH that was indistinguishable from hypothalamic LHRH. Furthermore, the irLHRH was apparently encoded by LHRH mRNA [44], and we have found that the sequences of hypothalamic and lymphocyte LHRH are identical [45]. The level of irLHRH, approximately 500 pg/20$\times 10^6$ lymphocytes, represented about 15% of the entire spleen content since 100–120$\times 10^6$ lymphocytes were obtained per splenic preparation. As such, the amount of irLHRH in spleen is quite comparable to that of single whole rat hypothalami.

Rat lymphocytes also synthesized and released relatively large amounts (200 pg/mg protein) of irGHRH as compared to hypothalamic cells (1,000 pg/mg protein) [46]. Once again, when we consider the greater protein content of the spleen relative to the hypothalamus, the potential RH output of these two organs is of the same order of magnitude. Northern gel analysis showed that lymphocyte GHRH-related RNA was polyadenylated and of the same molecular mass (0.8 kDa) as GHRH mRNA. Antibody affinity chromatography followed by size separation columns demonstrated two peaks of irGHRH. The smaller of these (5 kDa) was the same size as GHRH and was de novo synthesized since it could be intrinsically radiolabeled with ^{3}H-amino acids. The 5-kDa irGHRH bound the GHRH receptor and increased GH mRNA synthesis in pituitary cells. Collectively, these studies demonstrate the production of hypothalamic RHs by immunocytes. Since immunocytes also contain the pituitary hormones that are regulated by these hypothalamic releasing factors, the data suggest that the immune system may regulate itself by using peptides that are analogous to those of the hypothalamic pituitary axis. This idea is made particularly attractive by the observation that CRF can elicit lymphocyte ACTH production [47] and GHRH can elevate levels of lymphocyte GH [38].

Neuropeptides

In addition to the pituitary hormones and their releasing factors, a number of neuropeptides have been isolated from cells of the immune system. Vasoactive intestinal peptide (VIP) and somatostatin (SOM) have been immunologically detected in platelets, mononuclear leukocytes, mast cells and polymorphonuclear (PMN) leukocytes [48–53]. Although the amino acid sequences have not been completed, the immune-derived SOM_{28} appears to be larger than the authentic neuropeptide [51] and to have a similar but not identical amino acid composition [50]. VIPs from rat basophilic leukemia cells consist of VIP_{10-28} free acid and a mixture of amino terminally extended 'big' VIPs, which are apparently obtained from a novel prepro-VIP encoded by an alternately spliced mRNA [51].

High levels of neuropeptide tyrosine (NPY) and its mRNA have been found in rat peripheral blood cells, bone marrow, and spleen. A combination of in situ hybridization and immunocytochemistry was used to show that the NPY-like material was present in megakaryocytes (i.e. platelet precursors) [54]. Another cell type, macrophages, were recently shown to synthesize ir substance P and its mRNA [55]. Arginine vasopressin (AVP), oxytocin, and neurophysin are detectable in a lymphoid-associated organ, the thymus, although they are apparently not localized to lymphocytes [56, 57]. The apparent source of these ir peptides are thymic 'nurse' cells [58]. Thymus extracted oxytocin and neurophysin eluted in the same position as reference standards on Sephadex G-75. The authenticity of oxytocin was also confirmed by biological assay and HPLC analysis. irAVP coeluted with the authentic peptide on reverse-phase HPLC [57].

A number of other neuroendocrine peptides, including parathyroid hormone-related protein [59, 60], calcitonin gene-related peptide [51], and insulin-like growth factor I (IGF-I) [61–63], have also been shown to be synthesized by or associated with cells of the immune system. Thus when one considers the number of molecules so far identified, it seems likely that virtually all known neuroendocrine peptides or proteins will be found in the immune system.

New Proteins

In addition to known neuroendocrine peptides and proteins, a move is afoot to identify novel molecules that are associated with the neuroendocrine and immune systems. Two different approaches have been successfully employed. First, a neuroendocrine tissue, the pituitary gland, was evaluated for the presence of new immunoregulatory substances. This led to the discovery of a novel (63 kDa) protein, suppressin, which inhibited the growth of transformed and normal lymphoid and neuroendocrine cells, enhanced natural killer cell activity, and induced interferon [64–66]. Suppressin was subsequently shown also to be synthesized by lymphocytes. The second approach involved the use of selective hybridization and molecular cloning to isolate cDNA derived from genes expressed predominantly in the neuroendocrine and immune systems. Two such rat genes, named neuroimmune 1 and 2, were identified [67, 68]. Sequence analysis showed that the gene products were unique and hybridization studies suggest that neuroimmune 1 protein may play a role during early postnatal neural development. It is anticipated that these two approaches will eventually lead to the discovery of a number of new peptides and proteins that are involved in immune-neuroendocrine interactions.

Regulation of Leukocyte-Derived Peptide Hormones

With regard to those peptide hormones that have been shown to be inducibly synthesized by lymphocytes in response to immunostimulants, a possibly interesting pattern seems to emerge. That is, in many instances, a given stimulus results in a particular hormone (table 1). For instance, LPS, SEA, and MLR induce ACTH, TSH, and CG, respectively. If this pattern continues to be followed and expanded then in the future we may be able to assign certain aspects of the pathophysiology of infectious diseases and tumors to a particular hormone or group of hormones that are produced by the immune system. Thus, changes in adrenal function during bacterial and viral infections might be in part associated with leukocyte ACTH while

Table 2. Pituitary hormones and their hypothalamic releasing factors

Hypothalamic releasing factor	Hormone	Pituitary cell type	Lymphocyte response to releasing factor
Corticotropin releasing factor	Corticotropin	Corticotroph	Marked elevation of corticotropin
Growth hormone releasing hormone	Growth hormone	Somatotroph	Marginal elevation of growth hormone
Thyrotropin releasing hormone	Thyrotropin	Thyrotroph	Marked elevation of thyrotropin
Luteinizing hormone releasing hormone	Luteinizing hormone	Gonadotroph	Marked elevation of luteinizing hormone
Vasoactive intestinal peptide[1]	Prolactin	Lactotroph	Not tested

[1] Derived from lactotroph rather than hypothalamus.

changes in thyroid function might occur during cell-mediated immune responses and involved T cell production of TSH.

In addition to immunoregulation of leukocyte-derived hormones, there are also control elements that are similar to those in the hypothalamic pituitary axis. For instance, β-endorphin follows a circadian rhythm though it differs from the pituitary [69]. In the central axis, the anterior pituitary gland contains at least five different types of secretory cells with each cell type containing a minimum of one hormone formed in this gland. Production and secretion of a particular hormone by anterior pituitary cells is controlled by a specific hormone from the hypothalamus. These are termed hypothalamic releasing factors (table 2). Each of the pituitary and/or hypothalamic hormones in turn are usually controlled in a negative fashion by end products of the particular neuroendocrine cascade. Thus, glucocorticoid hormones, for example, suppress ACTH production and release. Cells of the immune system, in addition to responding to immunostimulants, also respond with fidelity to the classical hypothalamic regulators of anterior pituitary hormones (table 2). For instance, CRF was observed to cause the de novo synthesis and release of leukocyte-derived ACTH and β-endorphin in vitro [47] and in vivo [70]. While it occurred at about 10-fold higher concentrations, AVP (a less potent regulator of pituitary ACTH) alone was also observed to have intrinsic CRF activity on leukocytes. At concentrations that

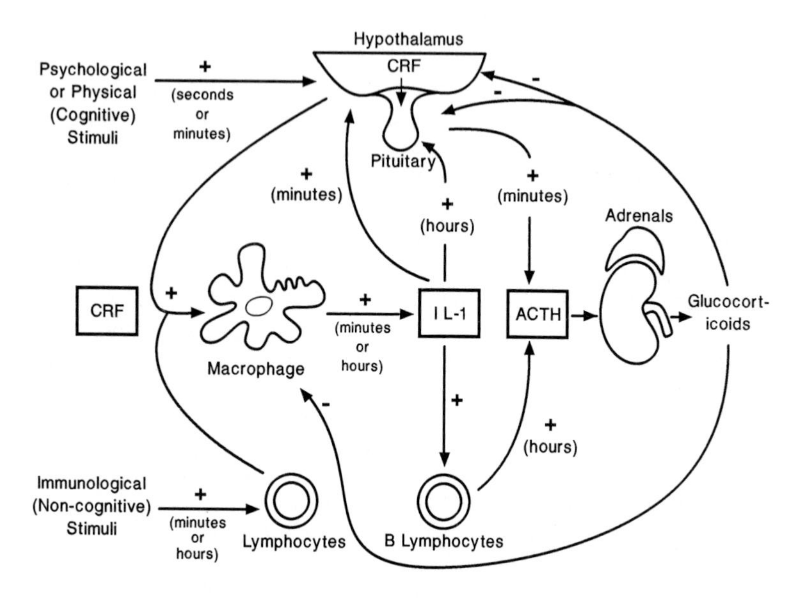

Fig. 1. Potential interplay of IL-1 and CRF in the lymphoid adrenal and pituitary adrenal axes. Cognitive or noncognitive stimuli are recognized by the nervous or immune systems, respectively, and hypothalamic or lymphoid CRF is released. CRF can in turn cause IL-1 release from macrophages or ACTH release from the pituitary gland. IL-1 can then elicit CRF from the hypothalamus and ACTH from the pituitary or B lymphocytes. ACTH then evokes a glucocorticoid response which can feedback inhibit IL-1 production from macrophages, CRF from the hypothalamus or ACTH from the pituitary. + = positive response; – = inhibitory response; ( ) = the time required for the response.

are frequently used on cultured pituitary cells, CRF and AVP together acted in an additive fashion to induce POMC-derived peptides and such induction was blocked by the synthetic glucocorticoid hormone, dexamethasone. Thus, on the surface, leukocytes seem quite similar to corticotrophs with respect to control of the POMC gene by CRF and AVP and feedback inhibition by a synthetic glucocorticoid hormone.

Upon closer examination, however, it was found that both the actions of CRF as well as dexamethasone were indirect (fig. 1). Heijnen and co-workers [71] have shown that CRF actually causes IL-1 production by macrophages and that the IL-1, rather than CRF, then elicits POMC production by B lymphocytes. The production of IL-1 in response to CRF is particularly interesting since the converse occurs in hypothalamic neurons. That is, IL-1 causes CRF release [72]. Glucocorticoids also acted indirectly by blocking

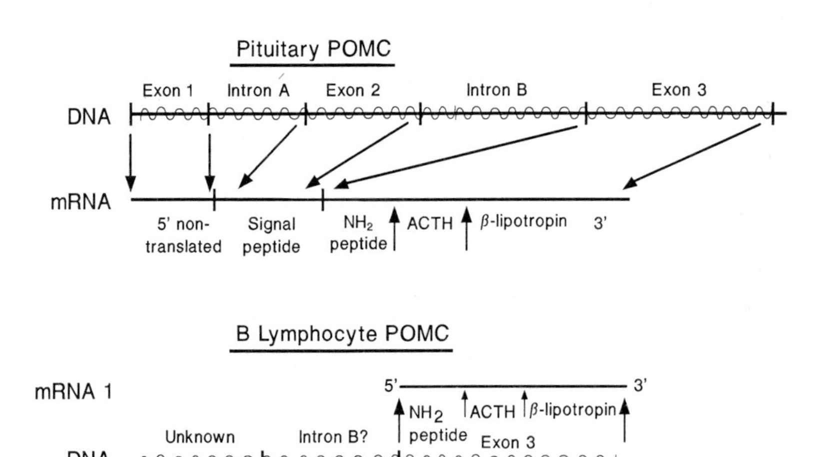

Fig. 2. Comparison of the structure of the pituitary POMC gene and a putative rearranged POMC gene in B lymphocytes. In the pituitary, part of the amino terminal peptide (NH_2) is in the 3′ end of exon 2. The larger B lymphocyte RNA would encode most of the NH_2 peptide and all of ACTH and β-lipotropin. The smaller mRNA would lack most of the NH_2 peptide but encode all of ACTH and β-lipotropin.

macrophage production of IL-1 instead of B cell POMC itself [73]. These findings suggest that the POMC gene in B lymphocytes is regulated differently from that in pituitary cells. In fact, we have recently shown that lymphocytes basally transcribe at least two POMC transcripts that are up-regulated by CRF presumably via IL-1 [15]. Of the 3 exons of POMC (fig. 2), the two lymphocyte mRNAs lacked exons 1 and 2 but contained either most or all of exon 3 which encodes most POMC-derived peptides (including ACTH) [14]. Since lymphocytes do not directly respond to CRF, we initially considered that this transcriptional difference resulted from the B lymphocyte use of an IL-1, rather than CRF, control element for the POMC gene. If this were the case, one would expect IL-1 to cause alternate transcription in pituitary cells. However, pituitary cells up-regulate only the expected 1-kb (3 exon) mRNA in response to either CRF or IL-1 indicating some other cause for the truncated mRNAs in lymphocytes [Galin, LeBoeuf, Blalock, unpubl. observation]. Based on these findings, it is tempting to speculate on the alternative explanation that some genetic rearrangement in proximity to the lymphocyte POMC gene leads to transcription initiation immediately 5′ to as well as within exon 3. Concomitant with this postulated rearrangement there is the apparent loss of the glucocorticoid regulatory element. Such rearrangement could easily be envisioned in the mouse B lymphocyte since the immunoglobulin heavy chain locus as well as the POMC gene are both located on chromosome 12.

Since lymphocytes have repeatedly been shown to produce POMC-derived peptides, such as ACTH and β-endorphin, at least some form of POMC message must be translated. Translation from a full-length, low-level POMC mRNA seems highly unlikely since POMC peptides are easily detected and the full-length mRNA was not detected even after extensive amplification by the polymerase chain reaction (PCR) [14, 15]. Therefore, translation of the truncated mRNAs would require the use of an alternate translation initiation site rather than the conventional site found within exon 2. The murine pituitary POMC exon 3 contains three possible AUG translation initiation codons. Each is in proper reading frame and are neighbored by 5′ and 3′ sequences which, though not optimal, would nonetheless allow translation [15]. However, only one of these AUG initiation codons resides 5′ to the ACTH coding region; the other two fall either within or 3′ to the ACTH coding sequence. Therefore, in order for ACTH production to occur in lymphocytes which has been conclusively demonstrated, the most 5′ translation initiation codon seems the best candidate. A final interesting problem is how the lymphocyte ACTH is secreted since exon 2 which encodes the signal sequence is missing. At present, we know that the ACTH or β-endorphin is extracellular [2, 3, 13, 70, 71, 74, 75] but do not know the mechanism for exit from the cell.

We have also observed a marginal up-regulation of T and B cell-derived irGH in response to GHRH [38]. Further similarity with the pituitary GH system is our recent finding that leukocyte GH can increase immunocyte synthesis of IGF-1 and that IGF-1 can in turn inhibit GH production by lymphocytes [63]. Similarly, thyrotropin-releasing hormone (TRH) has been found to induce the synthesis of TSH in mouse leukocytes and the human T cell leukemia cell line (MOLT-4) and such induction was blocked by T3 [18]. LHRH induction of lymphocyte LH is the most recent example of hypothalamic RF control of an immune system hormone [28–31, 34]. Thus, although there can be novel intermediates such as IL-1 involved, cells of the immune system seem quite similar to circulating pituitary cells with respect to their ability to positively respond to hypothalamic releasing factors and to have such responses blunted by the appropriate neuroendocrine end products of the particular hypothalamic pituitary axis (table 2).

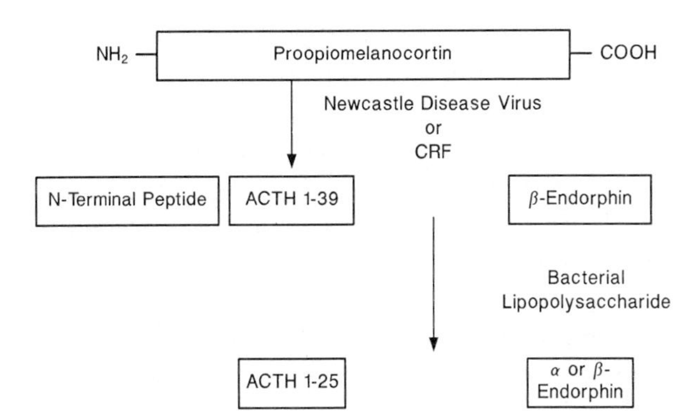

Fig. 3. Possible signal-dependent processing of lymphocyte POMC.

Processing of Leukocyte-Derived Peptide Hormones

Many peptide hormones and peptide neurotransmitters are synthesized as large prohormones that must be cleaved to release the active fragments. In the ACTH and endorphin system, this seems to result from cleavage of the precursor POMC at pairs of basic amino acids by a unique set of proteases. The site of POMC production can determine the final set of active peptides. For instance, in the anterior lobe of the pituitary gland, POMC is ultimately cleaved into an N-terminal fragment, ACTH (1-39) and β-lipotropin while the intermediate lobe ultimately yields γ-melanocyte-stimulating hormone (MSH), α-MSH, corticotropin-like intermediate lobe peptide (CLIP), γ-lipotropin and β-endorphin [76]. To date, pituitary proteases have been described which yield POMC-processing patterns which are similar, if not identical, to those seen in the anterior and intermediate lobes of the pituitary [77, 78]. The discovery of POMC production by cells of the immune system has led to the observation of processing pathways that are unique to leukocytes as well as composites of those observed in the anterior and intermediate lobe of the pituitary gland (fig. 3). With regard to previously described pituitary pathways, we have found that Newcastle disease virus (NDV) or CRF elicits the production of ACTH (1-39) and β-endorphin from cultured leukocytes [2, 47]. Others have shown that a subpopulation of mouse splenic macrophages spontaneously produce POMC which is processed into ACTH and β-endorphin [4]. In contrast, bacterial endotoxin or LPS treatment of leukocytes results in a novel processing pattern. It elicits the production of ACTH (1-25) [13] and endorphins which correspond to the molecular weight of α- or γ-endorphin from B lymphocytes [3, 79].

These findings point to some proteolytic cleavages of POMC as have been previously observed in the anterior and intermediate lobe of the pituitary as well as the hypothalamus. Of course, the results also suggest that cells of the immune system differ from virtually all other extrapituitary tissues where the major proteolytic cleavages are similar to those in the intermediate lobe of the pituitary gland [76]. For example, though we detect β-endorphin, we have yet to observe the production of an α-MSH-like peptide. Further, bacterial LPS induction of ACTH (1-25) suggests a quite novel processing pathway. In fact, we have recently demonstrated a B cell-derived enzyme which may be responsible for this alternate processing. This protease is induced or activated by LPS and at pH 5 cleaves ACTH (1-39) to a form that comigrates with ACTH (1-24) on polyacrylamide gels [80]. Such differential processing points to cells of the immune system having processing pathways which are both unique and in some instances composites of those seen in the anterior and intermediate lobe of the pituitary gland. A further interesting implication of this work is the suggestion of a possible stimulus-specific B lymphocyte-processing mechanism for POMC. This could be quite different from previously described pathways which are largely determined by the cell type (i.e. intermediate vs. anterior lobe pituitary cells) in which the POMC is processed. This could be related to the level of maturity of the B cell. Within the immune system, such differential processing could have important immunoregulatory consequences. For instance, α- but not β-endorphin suppresses in vitro antibody responses while β- but not α-endorphin enhances T cell mitogenesis [81, 82]. Also, ACTH (1-39) but not ACTH (1-24) suppresses in vitro antibody production [81]. Thus, the type of POMC stimulus may determine the processing pathway and ultimately the specific peptides which result. The specific peptide, in turn, then determines which immunologic cell type and function will be affected. Of course, the end result would be that different stimuli would elicit different responses via the same prohormone. Similarly, the isolation of distinct fragments of VIP (1-28) from lymphocytes, mast cells, and other leukocytes is another example of posttranslational peptidolysis [51]. In this sytem, the peptide variants bind differently to lymphocyte than to neural receptors, suggesting that each member of the VIP family may have a distinct immunoregulatory role [83].

In addition to having proteolytic processing enzymes, macrophages, at least, must also contain acetylating enzymes. This is based on the finding that while the major endorphin species in mouse spleen macrophages is β-endorphin (β-EP$_{1-31}$), there are smaller amounts of N-acetylated β-EP$_{1-16}$ (α-endorphin), β-EP$_{1-17}$ (γ-endorphin), β-EP$_{1-27}$ and β-EP$_{1-31}$ [10], While the study of regulatory factors which control leukocyte-derived hormone genes and enzymes which posttranslationally modify their products are just beginning, it is clear that this will be an important and exciting area of future investigation.

Immunologic Cell Types that Produce Peptide Hormones

While little work has been done on the production of peptide hormones by particular cell types within the immune system, a few observations are worthy of mention. Essentially, the cellular origin of the peptide depends on the particular hormone and the stimulus for production. For instance, when human peripheral blood cells are infected with NDV and all cells harbor the virus, then all cells produce POMC-derived peptides. Thus in response to this stimulus, every major cell type (T and B lymphocytes, NK cells, and macrophages) has the potential to produce POMC-derived peptides [1]. On the other hand, when these cells are treated with bacterial LPS only, B cells produce the peptides [79]. Since all of the cells have the potential to produce this group of hormones, this is apparently an instance where the stimulus is only recognized by B lymphocytes and hence only these cells produce ACTH and endorphins. In contrast, the production of TSH seems to be limited to T lymphocytes regardless of the stimulus. Thus, in this instance only, T cells seem to have the potential to make this hormone [17, 18]. In contrast to TSH, T and B cells as well as macrophages produce GH [40]. Interestingly, GH is primarily an enhancer of immune response and within the T cell compartment, T-helper cells produce more GH than T cytotoxic/suppressor cells. Obviously, a great deal of work remains to be done in this area. However, the preliminary results indicate there may be some cell specificity to certain hormonal responses.

Functions of Leukocyte-Derived Peptide Hormones

At least two possibilities exist relative to the function of peptide hormones and neurotransmitters that are produced by the immune system. First they act on their classic neuroendocrine target tissues. Second, they may serve as endogenous regulators of the immune system. With regard to the latter possibility, it is clear that neuroendocrine peptide hormones can directly modulate immune functions, and this is discussed in detail in another chapter of this book. These studies, however, do not address endogenous as opposed to exogenous regulation by neuroendocrine peptides. A number of investigators have now been able to demonstrate that such regulation is endogenous to the immune system. Specifically, TSH is a pituitary hormone that can be produced by lymphocytes in response to TRH and, like TSH, TRH enhanced the in vitro antibody response [84]. This enhancement was not observed with GHRH, AVP, or LHRH and was blocked by antibodies to the β subunit of TSH [19]. Thus, it appears that TRH specifically enhances the in vitro antibody response via production of TSH. This was apparently the first demonstration that a neuroendocrine hormone (TSH) can function as an endogenous regulator within the immune system. In a similar study, antibody to prolactin was shown to inhibit mitogenesis through neutralization of the lymphocyte-associated prolactin [85].

Two new and different approaches have provided further convincing evidence that endogenous neuroendocrine peptides have autocrine or paracrine immunoregulatory functions. First, an opiate antagonist was shown to indirectly block CRF enhancement of NK cell activity by inhibiting the action of immunocyte-derived opioid peptides [86]. Second, Weigent et al. [87] have used an antisense oligodeoxynucleotide (ODN) to the translation start site of GH mRNA to specifically inhibit leukocyte production of GH. The ensuing lack of GH resulted in a marked diminution in basal rates of DNA synthesis in such antisense ODN-treated leukocytes which was overcome by exogenously added GH. Certain experimental models and clinical observations would also seem to support the view that leukocyte-derived hormones can also act on their classic neuroendocrine targets. The finding that cells of the immune system are a source of secreted ACTH suggested that stimuli which elicit the leukocyte-derived hormone should not require a pituitary gland for an ACTH-mediated increase in corticosteroids. This seemed to be the case since NDV (an in vitro inducer of leukocyte ACTH) infection of hypophysectomized mice caused a time-dependent increase in corticosterone production which was inhibitable by dexamethasone. Unless such mice were pretreated with dexamethasone, their spleens were positive for ACTH by immunofluorescence [8]. A more recent study has strongly suggested that B lymphocytes can be responsible for extrapituitary ACTH and glucocorticoid production. In this report, hypophysectomized chickens were shown to produce ACTH and corticosterone in response to *Brucella abortus*. This ACTH and corticosterone response was ablated if B lymphocytes were deleted by bursectomy prior to hypophysectomy [88]. Similar results have been observed in humans. For example, when children who were pituitary ACTH-deficient were pyrogen tested (typhoid vaccine, another in vitro inducer of leukocyte ACTH), they showed an increase in the percentage of ACTH-positive mononuclear leukocytes [89]. Both the response in hypophysectomized mice and hypopituitary children peaked at approximately 6–8 h after administration of virus and typhoid vaccine, respectively. These findings might explain the earlier observation of bacterial polysaccharide (Piromen)-induced cortisol responses in 7 out of 8 patients who underwent pituitary stalk sectioning [90]. Such studies have been furthered by the report of Fehm et al. [91] that CRF administration to pituitary ACTH-deficient individuals results in both an ACTH and cortisol response.

Gram-negative bacterial infections and endotoxin shock may represent yet another situation in which leukocyte hormones act on the neuroendocrine system. For instance, endorphins have been implicated in the pathophysiology associated with these maladies since the opiate antagonist, naloxone, improved survival rates and blocked a number of cardiopulmonary changes associated with these conditions [92]. Further, two separate pools of endorphins have been observed following endotoxin, bacterial LPS, administration and it was suggested that one pool might originate in the immune system [93]. Considering the potent immunological effects of endotoxin, as well as its ability to induce in vitro leukocyte production of endorphins, cells of the immune system seem the most likely source of endogenous opiates that are observed during gram- negative sepsis and endotoxin shock. Consistent with this idea is the observation that lymphocyte depletion, like naloxone treatment, blocked a number of endotoxin-induced cardiopulmonary changes [94]. Our interpretation of these results is that lymphocyte depletion removes the source of the endorphins while naloxone blocks their effector function. In a different approach, LPS-resistant inbred mice which have essentially no pathophysiologic response to LPS were shown to have a defect in leukocyte processing of POMC to endorphins. If leukocyte-derived endorphins were administered to such LPS-resistant mice, they then showed much of the pathophysiology associated with LPS administration to sensitive mice [95]. A final and exciting new development in the opioid field has come with the demonstration that activation of endogenous opioids in rats by a cold water swim results in a local antinociceptive effect in inflamed peripheral tissue. This local antinociception in the inflamed tissue apparently results from production by immune cells of endogenous opioids which interact with opioid receptors on peripheral sensory nerves [96].

In conclusion, cells of the immune system can now be considered a source of neuroendocrine peptides. In many respects, the production and regulation of these peptides in leukocytes is remarkably like that observed in neuroendocrine cells. There are, however, a number of noteworthy differences which suggest that rules which apply to pituitary hormone production are not necessarily applicable to the immune system. First, there may not be an immediate response of the immune system since, unlike the pituitary gland, the hormones are usually not stored but are synthesized de novo and this requires a number of hours. Second, plasma hormone concentrations do not have to reach the levels required when the pituitary gland is the source because immune cells are not fixed but are mobile and can locally deposit the hormone at the target site. Third, on a per cell basis, leukocytes produce considerably less hormone than pituitary cells. This difference, however, is more than compensated for by the greater number of cells in the immune system as compared with the pituitary gland. Once produced, these peptide hormones seem to function in at least two capacities. They are endogenous regulators of the immune system as well as conveyors of information from the immune to the neuroendocrine system. It is our bias that the transmission of such information to the neuroendocrine system represents a sensory function for the immune system wherein leukocytes recognize stimuli that are not recognizable by the central and peripheral nervous systems [97]. These stimuli have been termed noncognitive and include bacteria, tumors, viruses, and antigens. The recognition of such noncognitive stimuli by immunocytes is then converted into information, in the form of peptide hormones and neurotransmitters, lymphokines and monokines, which is conveyed to the neuroendocrine system and a physiologic change occurs.

Acknowledgments

I thank my faculty colleagues, students and postdoctoral fellows who were pivotal to the studies reported herein. I am also grateful to Diane Weigent for expert editorial assistance. These studies were supported by a Cancer Center Core grant CA13148, an NIH grant DK38024, and a Council for Tobacco Research grant.

References

1 Blalock JE, Smith EM: Human leukocyte interferon: Structural and biological relatedness to adrenocorticotropic hormone and endorphins. Proc Natl Acad Sci USA 1980;77:5972–5974.
2 Smith EM, Blalock JE: Human lymphocyte production of ACTH and endorphin-like substances: Association with leukocyte interferon. Proc Natl Acad Sci USA 1981;78:7530–7534.
3 Harbour-McMenamin D, Smith EM, Blalock JE: Bacterial lipopolysaccharide induction of leukocyte derived ACTH and endorphins. Infect Immun 1985;48:813–817.
4 Lolait SJ, Lim ATW, Toh BH, Funder JW: Immunoreactive β-endorphin in a subpopulation of mouse spleen macrophages. J Clin Invest 1984;73:277–280.
5 Endo Y, Sakata T, Watanabe S: Identification of proopiomelanocortin-producing cells in the rat pyloric antrum and duodenum by in situ mRNA-cDNA hybridization. Biomed Res 1985;6:253–256.
6 Herbert E: Discovery of pro-opiomelanocortin – A cellular polyprotein. Trends Biochem Sci 1981;6:184–186.
7 Blalock JE, Smith EM: A complete regulatory loop between the immune and neuroendocrine systems. Fed Proc 1985;44:108–111.
8 Smith EM, Meyer WJ, Blalock JE: Virus-induced increases in corticosterone in hypophysectomized mice: A possible lymphoid adrenal axis. Science 1982;218:1311–1313.
9 Westley HJ, Kleiss AJ, Kelley KW, Wong PKY, Yuen P-H: Newcastle disease virus-infected splenocytes express the pro-opiomelanocortin gene. J Exp Med 1986;163:1589–1594.
10 Lolait SJ, Clements JA, Markwick AJ, Cheng C, McNally M, Smith AI, Funder JW: Pro-opiomelanocortin messenger ribonucleic acid and posttranslational processing of beta-endorphin in spleen macrophages. J Clin Invest 1986;77:1776–1779.
11 Oates EL, Allaway GP, Armstrong GR, Boyajian RA, Kehr JR, Prabhakar BS: Human lymphocytes produce pro-opiomelanocortin gene related transcripts. Effects of lymphotropic viruses. J Biol Chem 1988;263:10041–10044.
12 Buzzetti R, McLoughlin L, Lavender PM, Clark AJL, Rees LH: Expression of pro-opiomelanocortin gene and quantification of adrenocorticotropic hormone-like immunoreactivity in human normal peripheral mononuclear cells and lymphoid and myeloid malignancies. J Clin Invest 1989;83:733–737.
13 Smith EM, Galin FS, LeBouef RD, Coppenhaver DH, Harbour DV, Blalock JE: Nucleotide and amino acid sequence of lymphocyte-derived corticotropin: Endotoxin induction of a truncated peptide. Proc Natl Acad Sci USA 1990;87:1057–1060.
14 Galin FS, LeBoeuf RD, Blalock JE: A lymphocyte mRNA encodes the adrenocorticotropin/β-lipotropin region of the pro-opiomelanocortin gene. Prog Neuro Endocrin Immunol 1990;3:205–208.
15 Galin FS, LeBoeuf RD, Blalock JE: Corticotropin-releasing factor up-regulates expression of two truncated pro-opiomelanocortin transcripts in murine lymphocytes. J Neuroimmunol 1991;31:51–58.
16 Zurawski G, Benedik M, Kamp BJ, Abrams JS, Zurawski SM, Lee FD: Activation of mouse T-helper cells induces abundant pre-proenkephalin mRNA synthesis. Science 1986;232:772–775.

17 Smith EM, Phan M, Coppenhaver D, Kruger TE, Blalock JE: Human lymphocyte production of immunoreactive thyrotropin. Proc Natl Acad Sci USA 1983;80:6010–6013.
18 Harbour DV, Kruger TE, Coppenhaver D, Smith EM, Meyer WJ: Differential expression and regulation of thyrotropin (TSH) in T cell lines. Mol Cell Endocrinol 1989;64:229–241.
19 Kruger TE, Smith LR, Harbour DV, Blalock JE: Thyrotropin: An endogenous regulator of the in vitro immune response. J Immunol 1989;142:744–747.
20 Hiestand PC, Mekler P, Nordmann R, Grieder A, Permmongkol C: Prolactin as a modulator of lymphocyte responsiveness provides a possible mechanism of action for cyclosporin. Proc Natl Acad Sci USA 1986;83:2599–2603.
21 Montgomery DW, Zukoski CF, Shah NG, Buckley AR, Pacholczyk T, Russell DH: Concanavalin A-stimulated murine leukocytes produce a factor with prolactin-like bioactivity and immunoreactivity. Biochem Biophys Res Commun 1987;145:692–698.
22 Kenner JR, Holaday JW, Bernton EW, Smith PF: Prolactin-like protein in murine lymphocytes: morphological and biochemical evidence. Prog Neuro Endocrin Immunol 1990;3:188–195.
23 DiMattia GE, Gellersen B, Bohnet HG, Friesen HG: A human B-lymphoblastoid cell line produces prolactin. Endocrinology 1988;122:2508.
24 DiMattia GE, Gellersen B, Duckworth ML, Friesen HG: Human prolactin gene expression: The use of an alternative noncoding exon in decidua and the IM-9-P3 lymphoblast cell line. J Biol Chem 1990;2165:16412.
25 Friesen HG, DiMattia GE, Took CKL: Lymphoid tumor cells as models for studies of prolactin gene regulation and action. Prog Neuro Endocrin Immunol 1991;4:1–9.
26 Clevenger CV, Russell DH, Appasamy PM, Prystowsky MB: Regulation of interleukin-2 driven T-lymphocytes. Proc Natl Acad Sci USA 1990;87:6460.
27 Harbour-McMenamin D, Smith EM, Blalock JE: Production of lymphocyte derived chorionic gonadotropin in a mixed lymphocyte reaction. Proc Natl Acad Sci USA 1986;83:6834–6838.
28 Ebaugh MJ, Smith EM: Human lymphocyte production of immunoreactive luteinizing hormone (abstract). FASEB 1988;2:1642.
29 Smith EM, Ebaugh MJ: Luteogenic activity from human leukocytes. Ann NY Acad Sci 1990;594:492–493.
30 Blalock JE, Costa O: Immune neuroendocrine interactions: implications for reproductive physiology. Ann NY Acad Sci 1980;564:261–266.
31 Costa O, Mulchahey JJ, Blalock JE: Structure and function of luteinizing hormone releasing hormone (LHRH) receptors on lymphocytes. Prog Neuro Endocrin Immunol 1990;3:55–60.
32 Maier RM, Chew BP: Stimulation of progesterone secretion by blood monocytes and lymphocytes in the porcine. Theriogenology 1989;33:1045–1056.
33 Standaert FE, Chew BP, Wong TS: Influence of blood monocytes and lymphocytes on progesterone production by granulosa cells from small and large follicles in the pig. Am J Reprod Immunol 1990;22:49–55.
34 Standaert FE, Chew BP, Wong TS, Michal JJ: Porcine lymphocytes secrete factors in response to LHRH to stimulate progesterone production by granulosa cells in vitro (abstract). Biol Reprod 1990;42:75.

35 Gorospe WC, Kasson BG: Lymphokines from concanavalin-A-stimulated lymphocytes regulate rat granulosa cell steroidogenesis in vitro. Endocrinology 1989;123:2462–2471.

36 Kasson BG, Tuchel TL: Identification and characterization of a lymphocyte produced substance with gonadotropin-like activity; in Program of the 71st Annual Meeting of the Endocrine Society, Seattle 1989, abstr 210.

37 Weigent DA, Baxter JB, Wear WE, Smith LR, Bost KL, Blalock JE: Production of immunoreactive growth hormone by mononuclear leukocytes. FASEB J 1988;2:2812–2818.

38 Guarcello V, Weigent DA, Blalock JE: Growth hormone releasing hormone receptors on thymocytes and splenocytes from rats. Cell Immunol 1991;136:291–302.

39 Hattori N,Shimatsu A, Sugita M, Kumagai S, Imura H: Immunoreactive growth hormone (GH) secretion by human lymphocytes augmented by exogenous GH. Biochem Biophys Res Commun 1990;168:396–401.

40 Weigent DA, Blalock JE: The production of growth hormone by subpopulations of rat mononuclear leukocytes. Cell Immunol 1991;135:55–65.

41 Weigent DA. Riley JE, Galin FS, LeBoeuf RD, Blalock JE: Detection of growth hormone and growth hormone releasing hormone-related messenger RNA in rat leukocytes by the polymerase chain reaction. Proc Soc Exp Biol Med 1991;198:643–648.

42 Stephanou A, Jessop DS, Knight RA, Lightman SL: Corticotrophin-releasing factor-like immunoreactivity and mRNA in human leukocytes. Brain Behav Immun 1990; 4:67–73.

43 Emanuele NV, Emanuele MA, Tentler J, Kirsteins L, Azad N, Lawrence AM: Rat spleen lymphocytes contain an immunoreactive and bioactive luteinizing hormone-releasing hormone. Endocrinology 1990;126:2482–2486.

44 Azad N, Emanuele NV, Halloran MM, Tentler J, Kelley MR: Presence of luteinizing hormone-releasing hormone (LHRH) mRNA in spleen lymphocytes. Endocrinology 1991;128:1679–1681.

45 Maier CC, Marchetti B, LeBoeuf RD, Blalock JE: Hypothalamic and thymic LHRH mRNA sequences are identical (submitted).

46 Weigent DA, Blalock JE: Immunoreactive growth hormone-releasing hormone in rat leukocytes. J Neuroimmunol 1990;29:1–13.

47 Smith EM, Morrill AC, Meyer WJ, Blalock JE: Corticotropin releasing factor induction of leukocyte-derived immunoreactive ACTH and endorphins. Nature 1986;322:881–882.

48 Cutz E, Chan W, Track N, Goth A, Said S: Release of vasoactive intestinal peptide in mast cells by histamine liberators. Nature 1978;275:661–662.

49 Giachetti A, Goth A, Said SI: Vasoactive intestinal polypeptide (VIP) in rabbit platelets, and rat mast cells. Fed Proc 1978;37:657.

50 Goetzl EJ, Chernov-Rogan T, Cooke MP, Renold F, Payan FE: Endogenous somatostatin-like peptides of rat basophilic leukemic cells. J Immunol 1985;135:2707–2712.

51 Goetzl EJ, Grotmol T, VanDyke RW, Turck CW, Wershil B, Galli SJ, Sreedharan SP: Generation and recognition of vasoactive intestinal peptide by cells of the immune system. Ann NY Acad Sci 1990;594:34–44.

52 Lygren I, Revhaug A, Burhol PG, Giercksky E, Jenssen TG: Vasoactive intestinal peptide and somatostatin in leukocytes. Scand J Clin Lab Invest 1984;44:347–351.

53 O'Dorisio MS, O'Dorisio TM, Cataland S, Balcerzak SP: Vasoactive intestinal peptide as a biochemical marker for polymorphonuclear leukocytes. J Lab Clin Med 1980;96:666–670.

54 Ericsson A, Larhammar D, McIntyre KR, Persson H: A molecular genetic approach to the identification of genes expressed predominantly in the neuroendocrine and immune systems. Immunol Rev 1987;100:261–277.
55 Pascual DW, Bost KL: Substance P production by macrophage cell lines: a possible autocrine function for this neuropeptide. Immunology 1990;71:52.
56 Geenen V, Legros J-J, Franchimont P, Baudrihaye M, Defresne M-P, Bonever J: The neuroendocrine thymus: Coexistence of oxytocin and neurophysin in the human thymus. Science 1986;232:508–511.
57 Markwick AJ, Lolait SJ, Funder JW: Immunoreactive arginine vasopressin in the rat thymus. Endocrinology 1986;119:1690–1696.
58 Geenen V, Robert F, Martens H, Benhida A, DeGiovanni G, Defresne M-P, Boniver J, Legios J-J, Martial J, Franchimont P: Biosynthesis and paracrine (cryptocrine actions of 'self') neurohypophysial-related peptides in the thymus. Mol Cell Endocrinol 1991;76:C27–C31.
59 Adachi N, Yamaguchi K, Miyake Y, Honda S, Hagasak K, Akiyama Y, Adachi I, Abe K: Parathyroid hormone-related peptide is a possible autocrine growth inhibitor for lymphocytes. Biochem Biophys Res Commun 1990;166:1088–1094.
60 Fukumoto S, Matsumoto T, Watanabe T, Takhashi H, Miyoshi I, Ogata E: Secretion of parathyroid hormone-like activity from human T-cell lymphotropic virus type I-infected lymphocytes. Cancer Res 1989;49:3849–3852.
61 Rom WN, Basset P, Fells GA, Nukiwa T, Trapnell BC, Crystal RG: Alveolar macrophages release an insulin-like growth factor 1-type molecule. J Clin Invest 1988;82:1685–1693.
62 Merimee TJ, Grant MB, Broder CM, Cavalli-Sforza LL: Insulin-like growth factor secretion by human B-lymphocytes: A comparison of cells from normal and pygmy subjects. J Clin Endocrinol Metab 1989;69:978–984.
63 Baxter JB, Blalock JE, Weigent DA: Characterization of immunoreactive insulin-like growth factor-I from leukocytes and its regulation by growth hormone. Endocrinology 1991;129:1727–1734.
64 LeBoeuf RD, Burns JN, Bost KL, Blalock JE: Isolation, purification and partial characterization of suppressin, a novel inhibitor of cell proliferation. J Biol Chem 1990;265:158–165.
65 LeBoeuf RD, Carr DJJ, Green MM, Blalock JE: Cellular effects of suppressin: A biological response modifier of cells of the immune system. Prog Neuro Endocrin Immunol 1990;3:176–187.
66 Carr DJJ, Blalock JE, Green MM, LeBoeuf RD: Immunomodulatory characteristics of a novel antiproliferative protein, suppressin. J Neuroimmunol 1990;30:179–187.
67 Ericsson A, Barbany G, Friedman WJ, Persson H: Molecular cloning and characterization of genes predominantly expressed in the neuroendocrine and immune systems. Prog Neuro Endocrin Immunol 1991;4:26–41.
68 Ericsson A, Barbany G, Persson H: Regional and temporal distribution of the neuroimmune mRNA NI-1 in the rat central nervous system. Prog Neuro Endocrin Immunol 1991:4:42–55.
69 Sacerdote P, Ciciliato IA, Rubboli F, Panerai AE: Effects of psychoactive drugs on lymphocyte neuropeptides. Ann NY Acad Sci 1990;594:270–279.
70 Kavelaars A, Berkenbosch F, Croiset G, Ballieux RE, Heijnen CJ: Induction of beta-endorphin secretion by lymphocytes after subcutaneous administration of corticotropin-releasing factor. Endocrinology 1990;126:759–764.

71 Kavelaars A, Ballieux RE, Heijnen CJ: The role of interleukin-1 in the CRF- and AVP-induced secretion of beta-endorphin by human peripheral blood mononuclear cells. J Immunol 1989;142:2338–2342.
72 Schettini G: Interleukin-1 in the neuroendocrine system: From gene to function. Prog Neuro Endocrin Immunol 1990;3:157–166.
73 Kavelaars A, Ballieux RE, Heijnen CJ: Beta-endorphin secretion by human peripheral blood mononuclear cells: Regulation by glucocorticoids. Life Sci 1990;46:1233–1240.
74 Buzzetti R, Valente L, Barietta C, Scavo D, Pozzilli P: Thymopentin induces release of ACTH-like immunoreactivity by human lymphocytes. J Clin Lab Immunol 1989; 29:157–159.
75 Clarke BL, Blalock JE, Gebhardt B: Mitogen-stimulated lymphocytes secrete biologically active corticotropin in a co-culture system. FASEB J 1991;A1486, abstr 6472.
76 Krieger DT: Brain peptides: What, where and why? Science 1983;222:975–985.
77 Loh YP, Parish DC, Tuteja R: Purification and characterization of a paired basic residue-specific pro-opiomelanocortin converting enzyme from bovine pituitary intermediate lobe secretory vesicles. J Biol Chem 1985;260:7194–7205.
78 Cromlish JA, Seidah NG, Chretien M: Selective cleavage of human ACTH, beta-lipotropin, and the N-terminal glycopeptide at pairs of basic residues by IRCM-serine protease. I. Subcellular localization in small and large vesicles. J Biol Chem 1986;261:10859–10870.
79 Harbour-McMenamin D, Smith EM, Blalock JE: B lymphocyte-enriched populations produce immunoreactive endorphin that binds to delta opiate receptors and may play a role in endotoxic shock. ICSU short reports. Adv Gene Technol: Mol Biol Endocr Syst 1986;4:378–379.
80 Harbour DV, Smith EM, Blalock JE: A novel processing pathway for proopiomelanocortin in lymphocytes: Endotoxin induction of a new pro-hormone cleaving enzyme. J Neurosci Res 1987;18:95–101.
81 Johnson HM, Smith EM, Torres BA, Blalock JE: Neuroendocrine hormone regulation of in vitro antibody production. Proc Natl Acad Sci USA 1982;79:4171–4174.
82 Gilman SC, Schwartz JM, Milner RJ, Bloom FE, Feldman JD: β-Endorphin enhances lymphocyte proliferative responses. Proc Natl Acad Sci USA 1982;79:4226–4230.
83 Wenger GD, O'Dorisio MS, Goetzl EJ: Vasoactive intestinal peptide: Messenger in a neuroimmune axis. Ann NY Acad Sci 1990;594:104–119.
84 Blalock JE, Johnson HM, Smith EM, Torres BA: Enhancement of the in vitro antibody response by thyrotropin. Biochem Biophys Res Commun 1985;125:30–34.
85 Hartmann DP, Holaday JW, Bernton EW: Inhibition of lymphocyte proliferation by antibodies to prolactin. FASEB J 1989;3:2194–2202.
86 Carr DJJ, DeCosta BR, Jacobson AE, Rice KC, Blalock JE: Corticotropin-releasing hormone augments natural killer cell activity through a naloxone-sensitive pathway. J Neuroimmunol 1990;28:53–61.
87 Weigent DA, Blalock JE, LeBoeuf RD: An antisense oligodeoxynucleotide to growth hormone messenger ribonucleic acid inhibits lymphocyte proliferation. Endocrinology 1991;128:2053–2057.
88 Bayle JE, Guellati M, Ibos F, Roux J: *Brucella abortus* antigen stimulates the pituitary-adrenal axis through the extra-pituitary B-lymphoid system. Prog Neuro Endocrin Immunol 1991;4:99–105.

89 Meyer WJ III, Smith EM, Richards GE, Cavallo A, Morrill AC, Blalock JE: In vivo immunoreactive ACTH production by human leukocytes from normal and ACTH-deficient individuals. J Clin Endocrinol Metabol 1987;64:98–105.
90 Van Wyk JJ, Dugger GS, Newsome JF, Thomas PZ: The effect of pituitary stalk section in the adrenal function of women with cancer of the breast. J Clin Endocrinol Metab 1960;20:157–172.
91 Fehm HL, Holl R, Spath-Schwalbe E, Voigt KH, Born J: Ability of human corticotropin releasing factor (hCRF) to stimulate cortisol secretion independent from pituitary ACTH. Life Sci 1988;42:679–686.
92 Reynolds DG, Guill NV, Vargish T, Hechner RB, Fader AI, Holaday JW: Blockade of opiate receptors for naloxone improves survival and cardiac performance in canine endotoxic shock. Circ Shock 1980;7:39–48.
93 Carr DB, Bergland R, Hamilton A, Blume H, Kasting N, Arnold M, Martin MB, Rosenblatt M: Endotoxin-stimulated opioid peptide secretion: Two secretory pools and feedback control in vivo. Science 1982;217:845–848.
94 Bohs CT, Fish JC, Miller TH, Traber DL: Pulmonary vascular response to endotoxin in normal and lymphocyte depleted sheep. Circ Shock 1979;6:13–21.
95 Harbour DV, Smith EM, Blalock JE: Splenic lymphocyte production of an endorphin during endotoxic shock. Brain Behav Immun 1987;1:123–133.
96 Stein C, Hassan AHS, Przewlocki R, Gramsch C, Peter K, Herz A: Opioids from immunocytes interact with receptors on sensory nerves to inhibit nociception in inflammation. Proc Natl Acad Sci USA 1990;87:5935–5939.
97 Blalock JE: The immune system as a sensory organ. J Immunol 1984;132:1067–1070.

Prof. J. Edwin Blalock, PhD, Department of Physiology and Biophysics,
University of Alabama at Birmingham, AL 35294–0005 (USA)

Blalock JE (ed): Neuroimmunoendocrinology, 2nd rev ed.
Chem Immunol. Basel, Karger, 1992, vol 52, pp 25–48

Noradrenergic and Peptidergic Innervation of Lymphoid Organs[1]

Suzanne Y. Felten, David L. Felten, Denise L. Bellinger, John A. Olschowka

Department of Neurobiology and Anatomy, University of Rochester School of Medicine, Rochester, N.Y., USA

Introduction

Anatomical studies from our laboratories [3, 9, 11, 12, 28, 30, 31, 34, 36, 38–43, 61, 82, 100, 101] and other laboratories [16, 17, 50, 58, 77, 91, 97, 98, 102] have revealed the presence of autonomic nerve fibers, mainly noradrenergic (tyrosine hydroxylase-positive sympathetic fibers), in both primary and secondary lymphoid organs, as well as mucosal associated lymphoid tissue. These fibers distribute into specific compartments of these organs (fig. 1–4), and are associated with both smooth muscle and specific cellular elements of the immune system, including lymphocytes, macrophages, and associated cells such as mast cells. Norepinephrine (NE), the postganglionic sympathetic neurotransmitter present in these fibers, appears to be available for secretion immediately adjacent to lymphoid cells with direct neuroeffector contacts on lymphocyte, such as those in the periarteriolar lymphatic sheath (PALS) [41, 42, 44], and for more general paracrine-like diffusion through widespread regions of the lymphoid organs. In view of past evidence for adrenergic receptors on lymphocytes and other lymphoid cells, and a functional role for NE in modulation of immune function, we propose that NE fulfills the criteria for neurotransmission with cells of the immune system as the target tissue. Evidence for this hypothesis is described briefly below, and is summarized elsewhere [7, 27, 32–34, 69].

NE is not the only neurotransmitter that influences immune responses. Neuropeptides, including vasoactive intestinal polypeptide (VIP), somato-

[1] Supported by grant R37 MH42076, RO1 NS25233, RO1 NS24761, P50 MH40381, and R29 MH47783. We thank John Housel for excellent technical assistance.

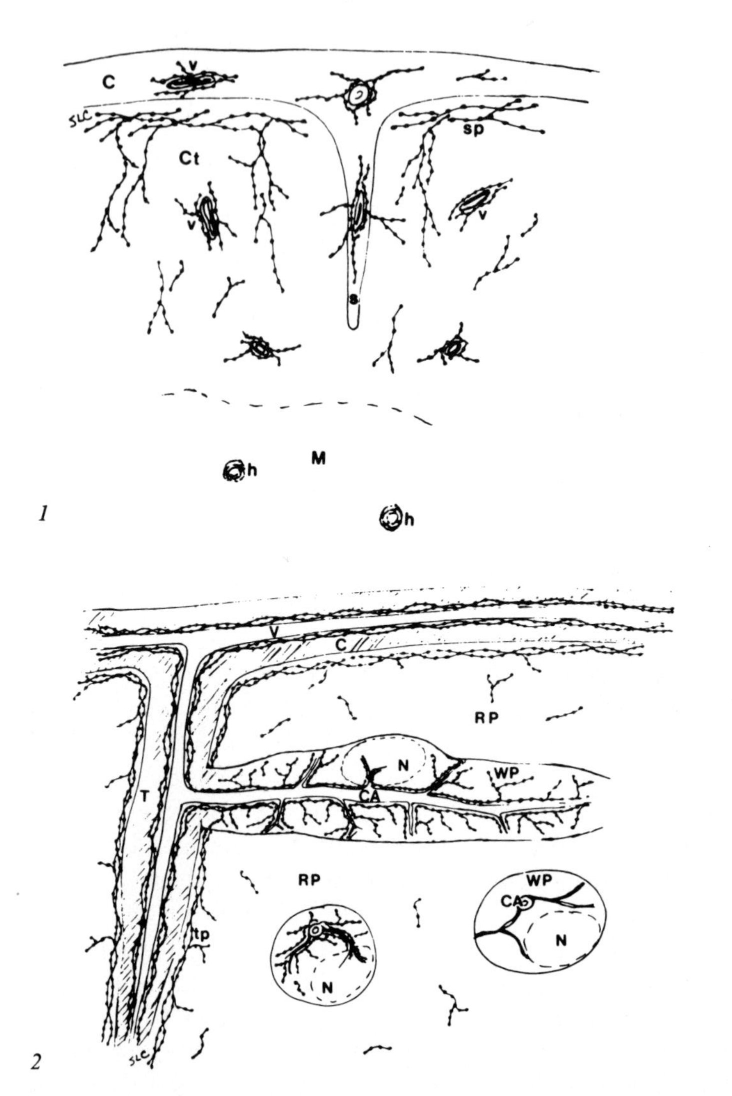

Fig. 1. Schematic drawing of noradrenergic innervation of the thymus. Varicose fibers with blood vessels and in the capsule (C) and associated septa (s) form a subcapsular plexus (sp) and give rise to scattered varicosities in the cortex (Ct), associated with blood vessels (v) and found as free profiles in the parenchyma. The medullary region (M) and Hassall's corpuscles (h) have only sparse innervation [From 61, with permission].

Fig. 2. Schematic drawing of noradrenergic innervation of the spleen. Varicose fibers enter with the splenic artery and its branches (V), continue with those vessels and with the capsule (C) into trabeculae (T), form a subcapsular plexus and associated trabecular-venous plexuses (tp), and follow both the vascular and trabecular plexuses into the white pulp (WP) with the central artery (CA) and its branches. Fibers diverge from this arterial

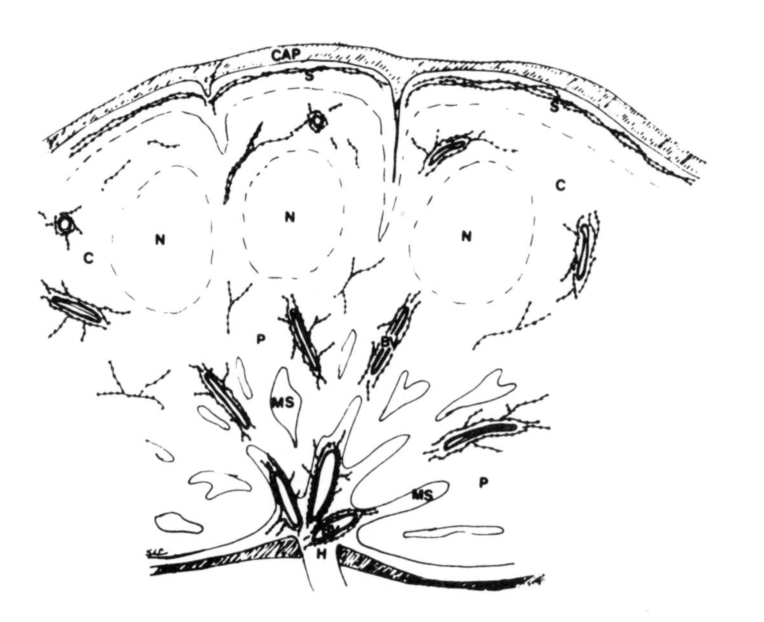

Fig. 3. Schematic drawing of noradrenergic innervation of a lymph node. Varicose fibers enter the node in the hilar region (H), travel with the vasculature (BV) in the medullary cords alongside the medullary sinuses (MS), and branch into the paracortical regions (P) adjacent to the nodules (N). Some varicosities form a subcapsular (S) plexus and give rise to fibers that travel into cortical (C) regions with blood vessels and as free profiles. Noradrenergic fibers are not found in nodular regions. CAP = capsule [From 61, with permission].

statin, substance P (SP), opioid peptides, and others can modulate some aspects of immune function through specific receptors on lymphocytes and other cells of the immune system [19, 51, 72, 78, 94]. Several neuropeptides, such as SP, somatostatin, VIP, calcitonin gene-related peptide (CGRP), neuropeptide Y (NPY) and opioid peptides, have been identified in nerves in

---

plexus into the white pulp and end among lymphocytes and macrophages in the parenchyma. Only scattered varicosities are found in the red pulp (RP) besides those in trabeculae, and noradrenergic fibers also avoid nodular regions (N) of the white pulp. The white pulp is represented in longitudinal section in the main schematic drawing. Cross sections of the white pulp are included below for distribution of the central arteriolar system (right) and its noradrenergic innervation (left) [From 61, with permission].

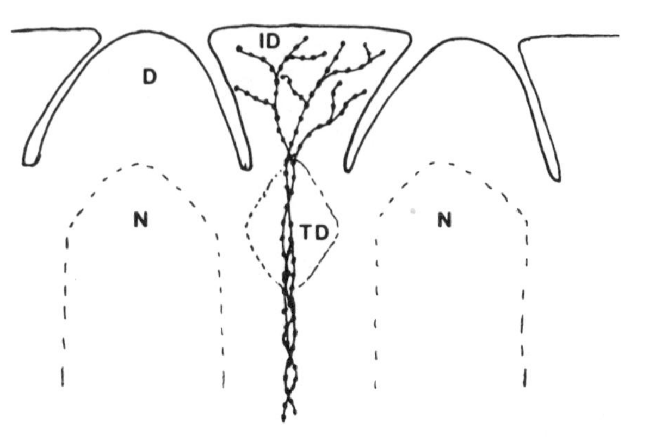

Fig. 4. Schematic drawing of noradrenergic innervation of GALT, represented by rabbit appendix. Noradrenergic fibers enter with the vasculature at the serosal surface, travel with Auerbach's plexus and along the muscularis interna, turn radially and travel in a tortuous internodular plexus between the large nodules (N), plunge through the T-dependent zone (TD) and arborize profusely in the interdomal region (ID) in the lamina propria. Here, the varicosities end among lymphocytes, enterochromaffin cells, and the subepithelial immunoglobulin-secreting plasma cells. The interdomal regions are separated from the adjacent domes (D) by a moat in continuity with the lumen of the gut [From 61, with permission].

spleen, lymph nodes, thymus, and other lymphoid tissue [13, 31, 36, 43, 47, 58, 59, 63, 64, 75, 77, 79, 80, 88, 91, 95, 98, 99, 102]. These neuropeptide-containing nerve fibers have distinct patterns of distribution and termination, often in association with cells of the immune system. Some neuropeptides appear to be co-localized with each other (e.g. SP and CGRP) or with a classical amine (e.g. NPY and NE), but these relationships are not invariable, and may differ from species to species, organ to organ, and compartment to compartment. Since lymphocytes, macrophages, and other cells of the immune system also possess receptors for many of these neuropeptides, and respond functionally to the neuropeptides, it is likely that they also can be viewed as neurotransmitters with cells of the immune system as targets [see 27 for detailed discussion of these criteria and how well they are fulfilled].

The presence of neurotransmitters in nerve fibers adjacent to cells of the immune system in lymphoid organs establishes an anatomical link that may provide a major route for translating central nervous system (CNS) activity into signals that can influence immune function. A large body of evidence has accumulated during the past decade documenting environmental and psychosocial factors that can influence immune functions through the CNS [reviewed in detail in 5, 7]. Immune responses can be enhanced or suppressed by classical conditioning. Numerous stressors, including bereavement from loss of a spouse, depression, confinement to a nursing home, or examination pressures in medical students, can lead to both altered measures of immune responses and altered health status, presumably as a consequence of the altered immune status, although definitive evidence for such a causal link is not yet available. Stressors in animals also can alter immune responses. Brain lesions in specific autonomic sites of the CNS can result in altered immune responses in either direction, depending upon the specific site. Thus, it appears that the CNS is capable of sending signals to the immune system that result in altered responses. The two available routes by which the CNS could signal the immune system include the neuroendocrine route and the autonomic nerves. Much attention has been focused on the neuroendocrine route in light of the immunological effects (largely immunosuppression) of glucocorticoid administration. However, not all CNS effects on the immune system can be explained by these steroids. Studies of both conditioned immunosuppression [6] and stress-induced lymphocyte hyporesponsiveness [56] have shown the continued presence of these responses in adrenalectomized animals. Many other hormones also have been shown to influence immune responses [14, 51, 52, 57, 70, 71, 76]. Central effects of intracerebroventricular CRF [54] and IL-1 [89] administration, and detailed mechanistic evaluation of stress-related changes in cellular immune responses [82], suggest that neural pathways mediate functional effects of CNS outflow. This evidence, coupled with the abundant presence of noradrenergic and peptidergic innervation of lymphoid organs, and appropriate adrenergic or peptidergic receptors on cells of the immune system, has focused our attention on these nerve fibers as a major communication link between the CNS and the immune system.

General Patterns of Innervation of Lymphoid Organs

Noradrenergic postganglionic sympathetic nerve fibers innervate both primary (bone marrow, thymus) (fig. 1) and secondary (spleen, lymph nodes, mucosal-associated lymphoid tissue (MALT)) (fig. 2–4) lymphoid organs. The anatomical connections appear to follow classical patterns from the spinal cord through autonomic sympathetic ganglia. The preganglionic cell bodies are found in the intermediolateral cell column of the spinal cord (e.g. T6-T12 level in rats for the splenic preganglionics), while the ganglion cells are found either in the sympathetic chain or in collateral ganglia (e.g. superior mesenteric-celiac ganglion for splenic postganglionics, superior cervical ganglion and other upper chain ganglia for thymic postganglionics, mesenteric ganglia for gut-associated lymphoid tissue (GALT) postganglionics), thus maintaining the classical two neuron connection from the CNS to the target tissue. The small, unmyelinated noradrenergic fibers generally enter the lymphoid organs with the vasculature, continue into the organ with the vasculature and the capsular/trabecular or septal system, and then arborize into the parenchyma (fig. 5–24) where they form close contacts with lymphocytes and macrophages (fig. 25–29). Our recent observations on the ontogeny of splenic noradrenergic fibers in the rat suggest that the directionality of branching is not necessarily from the vascular nerve plexuses into the parenchyma; indeed, it may be the other way around [4]. Perivascular 'guide fibers' are present in the spleen at the presumptive border of the marginal sinus before that sinus fully forms, before macrophages and lymphocytes distribute into specific compartments of the splenic white pulp, before smooth muscle cells are present along the arterioles, and before the vascular nerve plexuses have begun to form. Thus, our attention is drawn to the parenchymal elements of the lymphoid organs that are associated with these fibers. The patterns of distribution are confined to specific compartments of the immune organs. In general, the noradrenergic fibers arborize into T lymphocyte zones and macrophage zones of the adult organs, and avoid the follicular or nodular zones of some B lymphocytes. However, this generalization may require modification in ontogeny; studies of noradrenergic fibers in the developing rat spleen show a close association of B lymphocytes with noradrenergic fibers prior to the development of follicles. Therefore, it has become clear that each lymphoid organ must be studied individually for the patterns of distribution of nerve fibers and the possible functions of those fibers, and that the stage of development or maturation can influence the pattern and the extent of this innervation. As further evidence of this individuality of distribution, noradrenergic nerve fibers that innervate the spleen and lymph nodes [1, 9, 10, 29, 35, 37, 40] but not the thymus [11], diminish with age, in parallel with the diminution of T-dependent immune function. Such diminution of peptidergic innervation, other than co-localized NPY in the spleen, has not been seen.

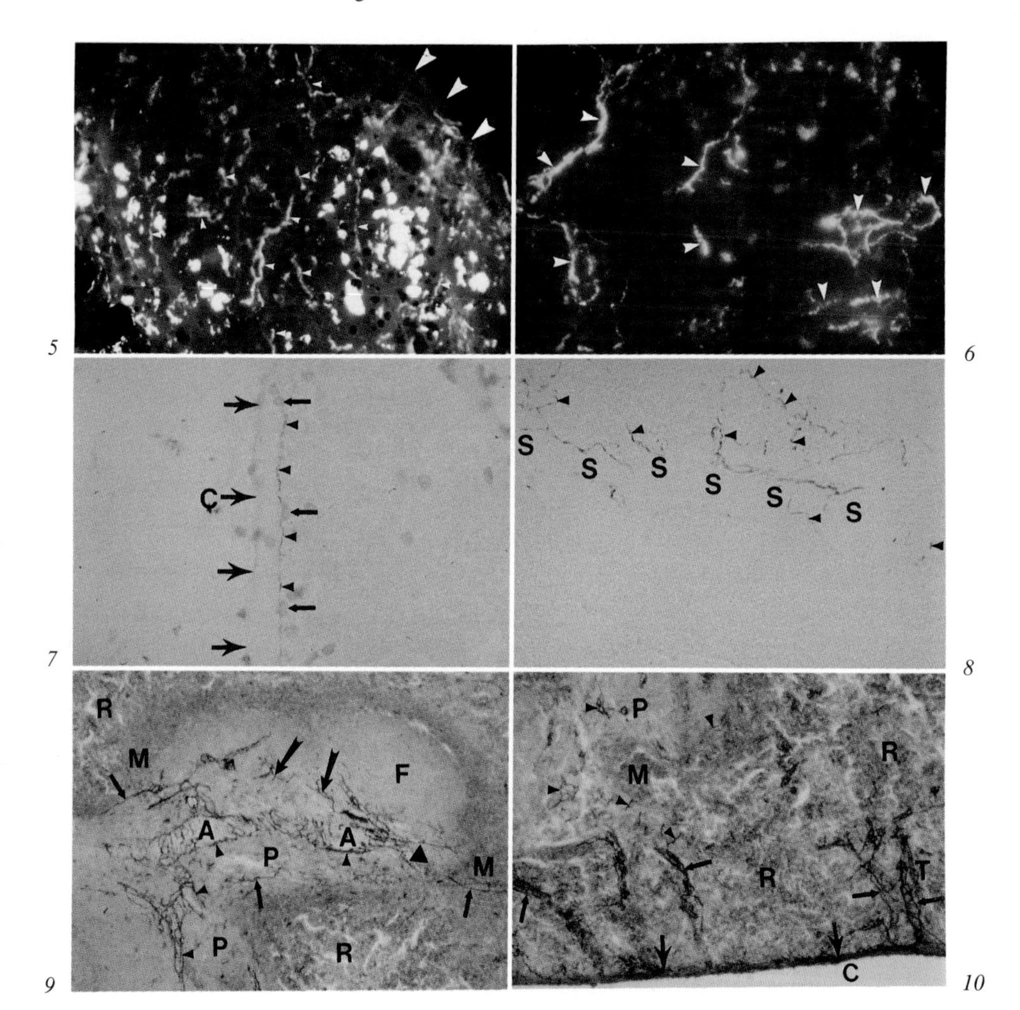

Fig. 5. Noradrenergic fibers (small arrowheads) in the thymic cortex. Varicose fibers distribute throughout the cortex, with dense profiles present in the subcapsular region and deep cortical region near the junction with the medulla. The capsular surface is denoted with large arrowheads. Adult rat. Fluorescence histochemistry. × 120.

Fig. 6. Vascular plexuses of noradrenergic fibers in the thymic cortex (denoted by arrowheads). Some nerve fibers distribute into the cortex adjacent to these plexuses. Adult rat. Fluorescence histochemistry. × 240.

Fig. 7. SP nerve fiber (arrowheads) running in a subcapsular location. This long linear profile runs adjacent to mast cells (small arrows). The capsular surface (C) is denoted by large arrowheads. Adult rat. Immunohistochemistry. × 120.

(For further legends see reverse side.)

Fig. 8. SP nerve fibers running along a septal (S) region in the thymic cortex, and extending into cortical regions (arrowheads) among thymocytes. Adult rat. Immunohistochemistry. × 120.

Fig. 9. Peripheral nerve fibers, immunostained by the general nerve marker PGP 9.5, in the white pulp of the spleen. Nerve fibers are present along the vasculature (small arrowheads) of the central arteriolar system (A), in parafollicular sites (large arrowhead) adjacent to the follicles (F), along the marginal sinus and in the marginal zone (M) (small arrows), and within the parenchyma of the PALS (P) (large arrows). R = red pulp. Adult rat. Immunohistochemistry. × 24.

Fig. 10. Peripheral nerve fibers, immunostained by the general nerve marker PGP 9.5, in the capsular/trabecular system and the white pulp of the spleen. Nerve fibers are present along the capsule (C), denoted by large arrows, and within the trabeculae (small arrows) coursing through the red pulp (R). Fibers also are present in the marginal zone (M) and PALS (P) of the white pulp (arrowheads). Adult rat. Immunohistochemistry. × 24.

Fig. 11. Peripheral nerve fibers, immunostained by the general nerve marker PGP 9.5, in large trabeculae (arrows) coursing through the red pulp (R) of the spleen. These trabeculae converge along large venous sinuses. Adult rat. Immunohistochemistry. × 24.

Fig. 12. TH-positive nerve fibers in the splenic white pulp. These fibers are present in a plexus around the central artery (A), along the marginal sinus (large arrowheads), and in the marginal zone (M) (arrow). Additional nerve fibers course through the parenchyma (small arrowheads) in the PALS (P), among T lymphocytes. R = red pulp. 14-day-old rat. Immunohistochemistry. × 120.

Fig. 13. TH-positive nerve fibers in the splenic white pulp. The fibers form a tangled nerve plexus around the central artery (A) and its branches. Fine varicose profiles (arrowheads) course through the parenchyma among T lymphocytes in the PALS (P). A large vein (V) also is innervated abundantly. 14-day-old rat. Immunohistochemistry. × 120.

Fig. 14. TH-positive nerve fibers ending among T lymphocytes in the white pulp of the spleen. The nerve fibers, nickel-enhanced to appear black, travel with the central arteriolar system (A), distribute along the marginal sinus (arrows), and course through the parenchyma (arrowheads) in the PALS (P) among T lymphoytes, immunostained with brown for the general T cell marker OX-19. M = marginal zone. Adult rat. Double label immunohistochemistry. × 120.

Fig. 15. TH-positive nerve fibers, nickel-enhanced to appear black, in the splenic white pulp along the marginal sinus (S) (large arrowheads), where ED-3-positive macrophages, immunostained brown, are clustered. An additional small nerve fiber bundle (small arrowheads) is found in a parafollicular region. F = follicle; A = central artery; M = marginal zone. Young adult rat. Double label immunohistochemistry. × 120.

Fig. 16. NPY nerve fibers in the white pulp of the spleen. NPY fibers form a plexus around the central artery (A), distribute through the parenchyma (arrowheads) in the PALS, and travel among IL-1-beta-positive cells along the marginal sinus in the marginal zone (M) (arrows). Adult rat. Double label immunohistochemistry (nerve fibers – black; IL-1-beta cells – brown). × 120.

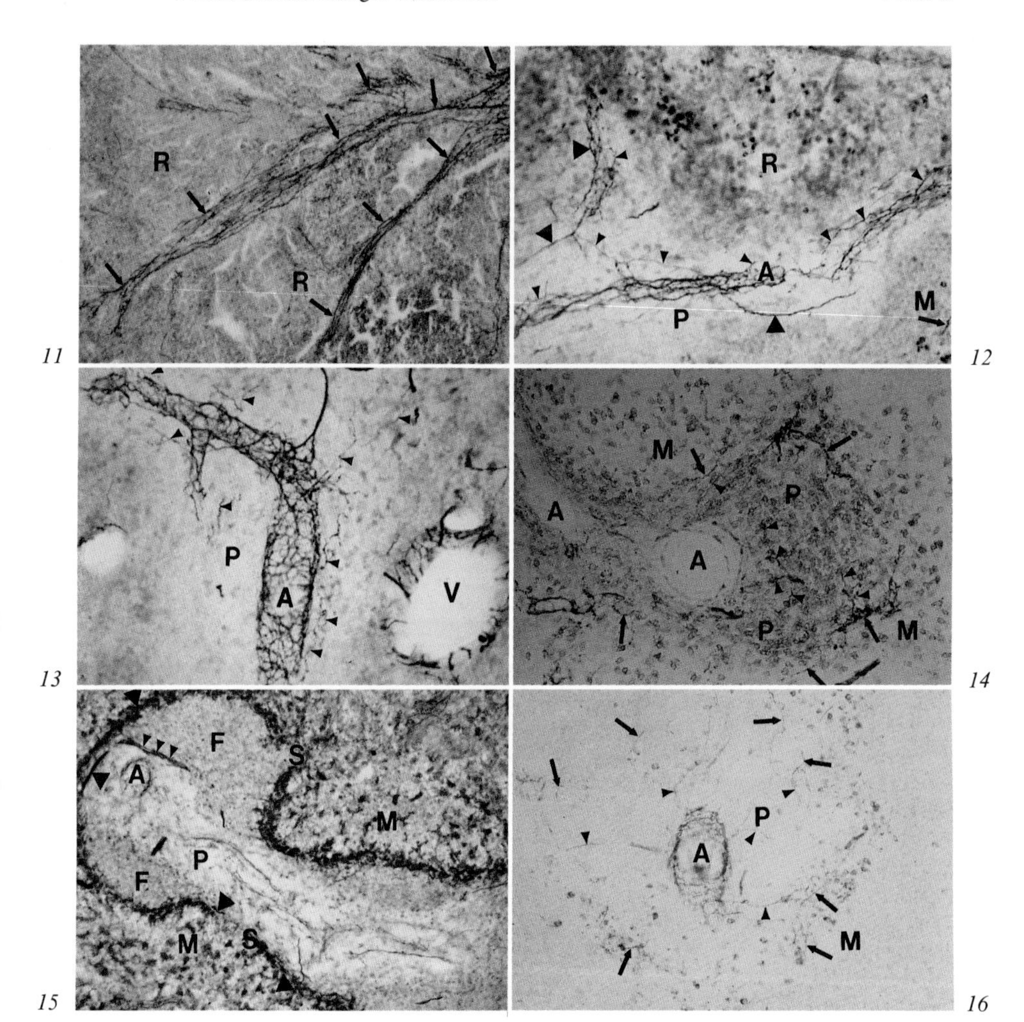
11
R
R
12
R
A
P
M
13
P
A
V
14
M
A
P
A
P
M
15
F
S
A
M
P
F
M
S
16
P
A
M

Fig. 17. SP nerve fibers (arrows) traveling along the venous sinuses (V) and in the red pulp (R) of the spleen. Adult rat. Immunohistochemistry. × 24.

Fig. 18. SP nerve fibers (arrows) traveling along the trabeculae and in the red pulp (R) of the spleen. Adult rat. Immunohistochemistry. × 60.

Fig. 19. A long linear SP nerve fiber (arrows) coursing through the white pulp (W) close to the boundary between the PALS and the marginal zone. The border between the white pulp and the red pulp (R) is denoted with arrowheads. Adult rat. Immunohistochemistry. × 60.

Fig. 20. Noradrenergic nerve fibers coursing through the medullary cords (MC) of a popliteal lymph node (large arrowheads), and distributing into the paracortical zone (P) (small arrowheads), where T lymphocytes are found. S = sinuses. Adult C3H mouse. Fluorescence histochemistry. × 60.

Fig. 21. Noradrenergic nerve fibers (arrowheads) scattered among lymphoid cells in the cortex (C) of a lymph node. Adult NZW mouse. Fluorescence histochemistry. × 120.

Fig. 22. NPY nerve fibers associated with the vascular system (arrows) and parenchymal regions (small arrowheads) of the medullary cords (MC) of a lymph node. S = sinuses. Adult rat. Immunohistochemistry. × 120.

Fig. 23. SP nerve fibers (arrows) traveling in the medullary cords (MC) along a sinus (S) and among T cytotoxic/suppressor lymphocytes. Adult rat. Double label immunohistochemistry (SP nerve fibers – black; OX-8-positive lymphocytes – brown). × 120.

Fig. 24. Noradrenergic nerve fibers (arrowheads) coursing through the lamina propria (LP) of the rabbit appendix. Yellow fluorescent enterochromaffin cells (arrows) are clustered in regions where these nerve fibers abound. Adult New Zealand white rabbit. Fluorescence histochemistry. × 120.

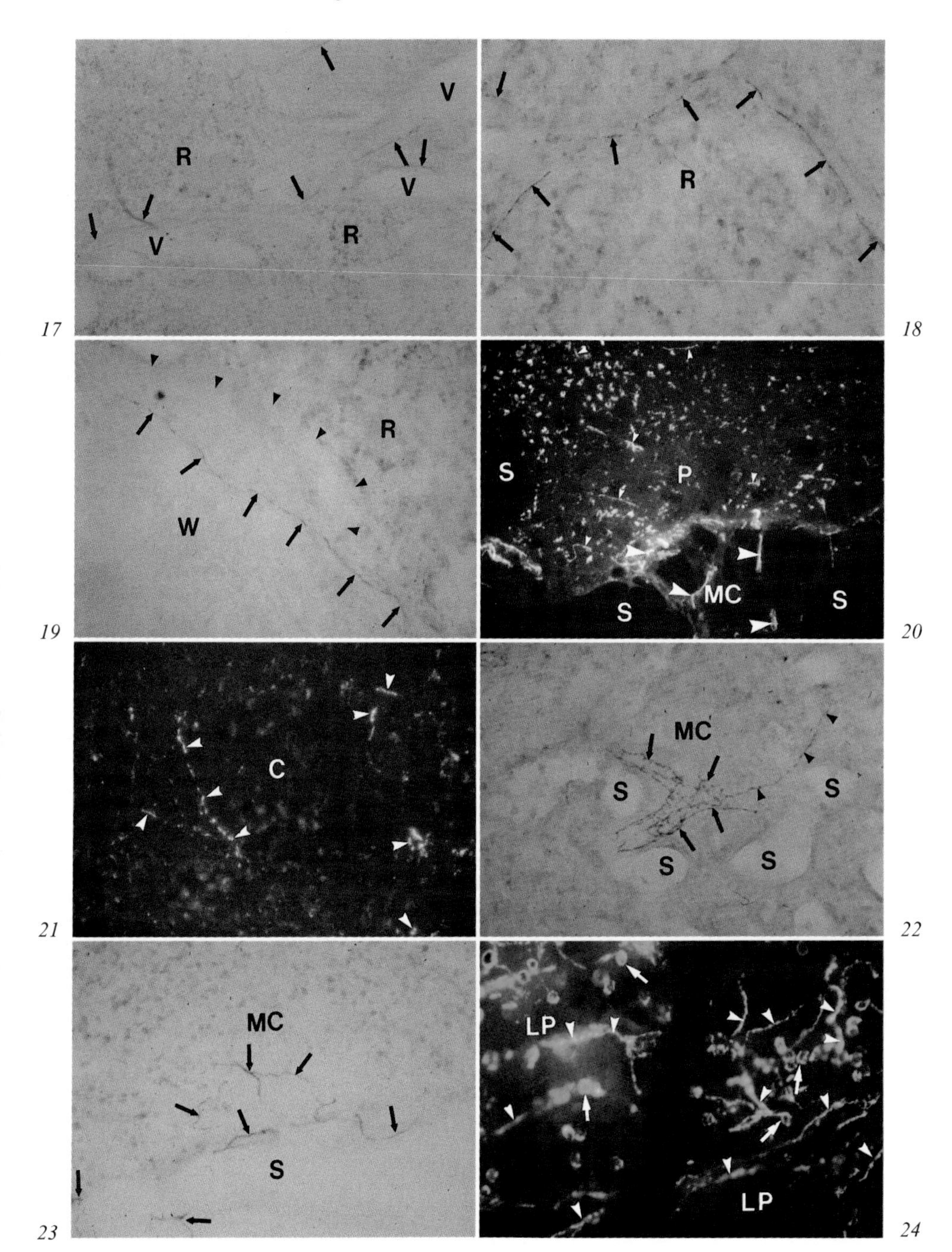
17
R
V
V
R
V
18
R
19
R
W
20
S
P
S
MC
S
21
C
22
MC
S
S
S
S
23
MC
S
24
LP
LP

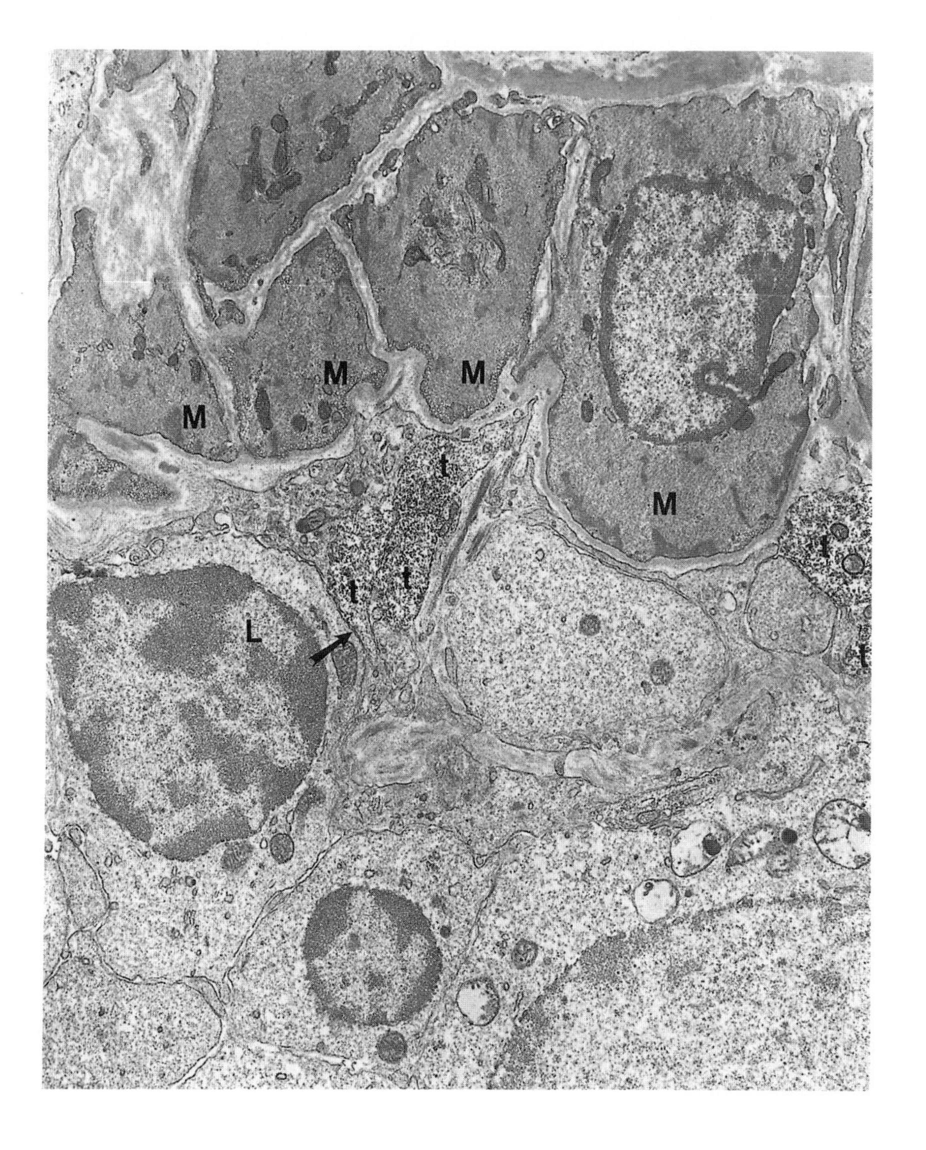

Fig. 25. TH-positive nerve terminals (t) containing the DAB reaction product in the adventitia of a central arteriole of the splenic white pulp. All of these terminals are separated from their adjacent smooth muscle cells (M) by a basement membrane and sometimes additional cell processes. However, a direct apposition is seen (arrow) between one TH-positive nerve terminal and a lymphocyte (L) in the PALS. Fischer 344 rat. EM immunohistochemistry. ×15,300.

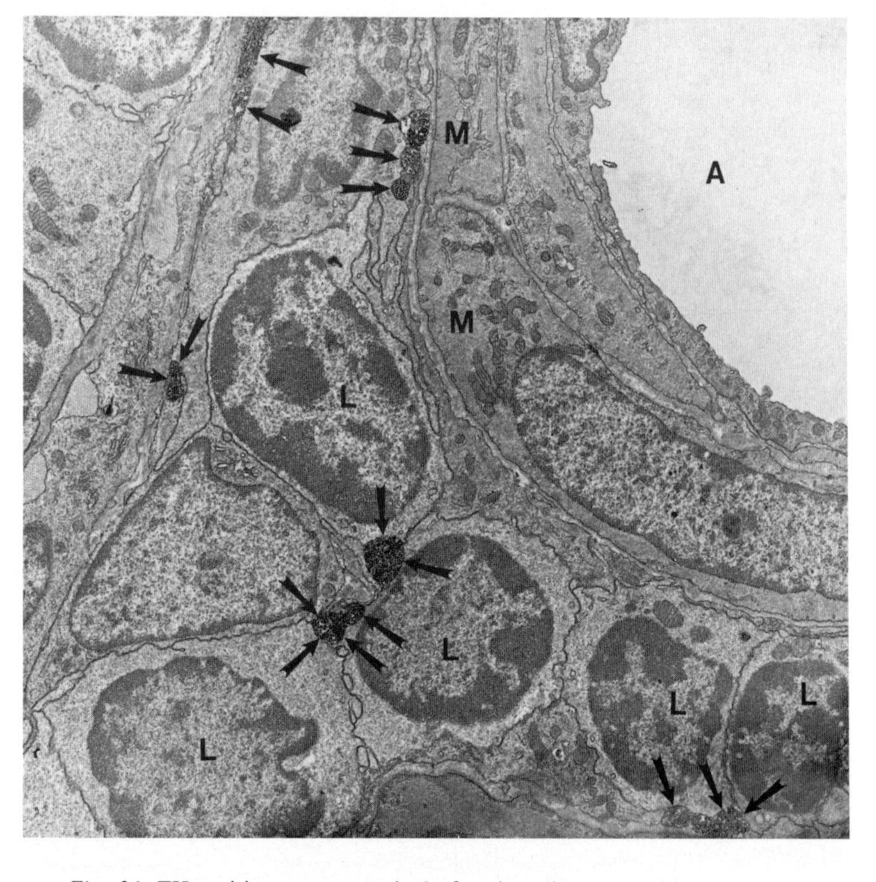

Fig. 26. TH-positive nerve terminals forming direct appositions (arrows) with lymphocytes (L) and other cells in the PALS of the white pulp. The three terminals in the upper center of the micrograph are located in the adventitia, separated from the smooth muscle cells (M) by a basement membrane, but directly abutting a cell on the distal side. The other TH-positive nerve terminals are found within the parenchyma, distant from the adventitia. A = lumen of the central arteriole. Fischer 344 rat. EM immunohistochemistry. ×15,300.*

Neuropeptide-containing fibers each possess their own distinct pattern of distribution in lymphoid organs. In general, NPY follows the pattern of distribution of NE, with which it is usually co-localized. SP and CGRP fibers often are very fine (perhaps sometimes below the range of detectable reaction product) and distribute with the vasculature (especially venous and trabecu-

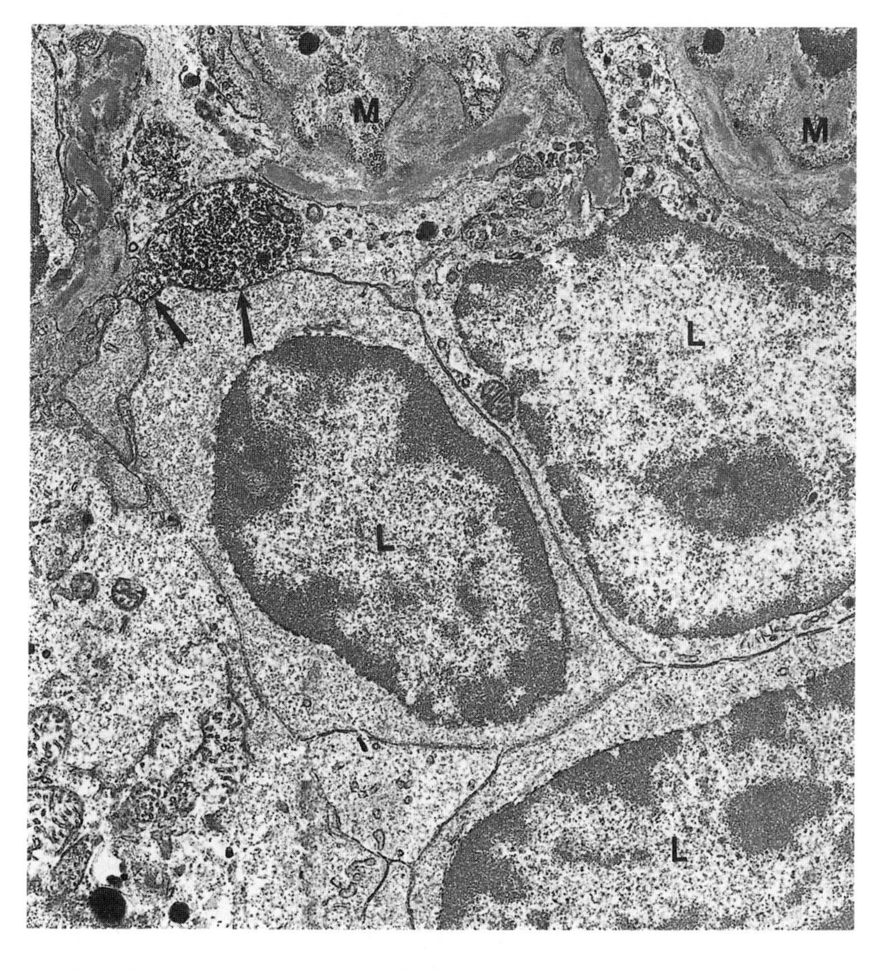

Fig. 27. TH-positive nerve terminals directly abutting (arrows) a lymphocyte (L) in the PALS of the white pulp. The larger terminal forms a smooth apposition, indented in the membrane of the lymphocyte. M = smooth muscle cell. Fischer 344 rat. EM immunohistochemistry. ×21,200.

lar systems) and among T cell zones, although other sites of localization such as red pulp of the spleen and the bursa of Fabricius have been described. The SP and CGRP fibers also are found closely abutting mast cells. VIP fibers also closely abut mast cells in the thymus. However complex the pattern of distribution is in each lymphoid organ, it is clear that neuropeptide-contain-

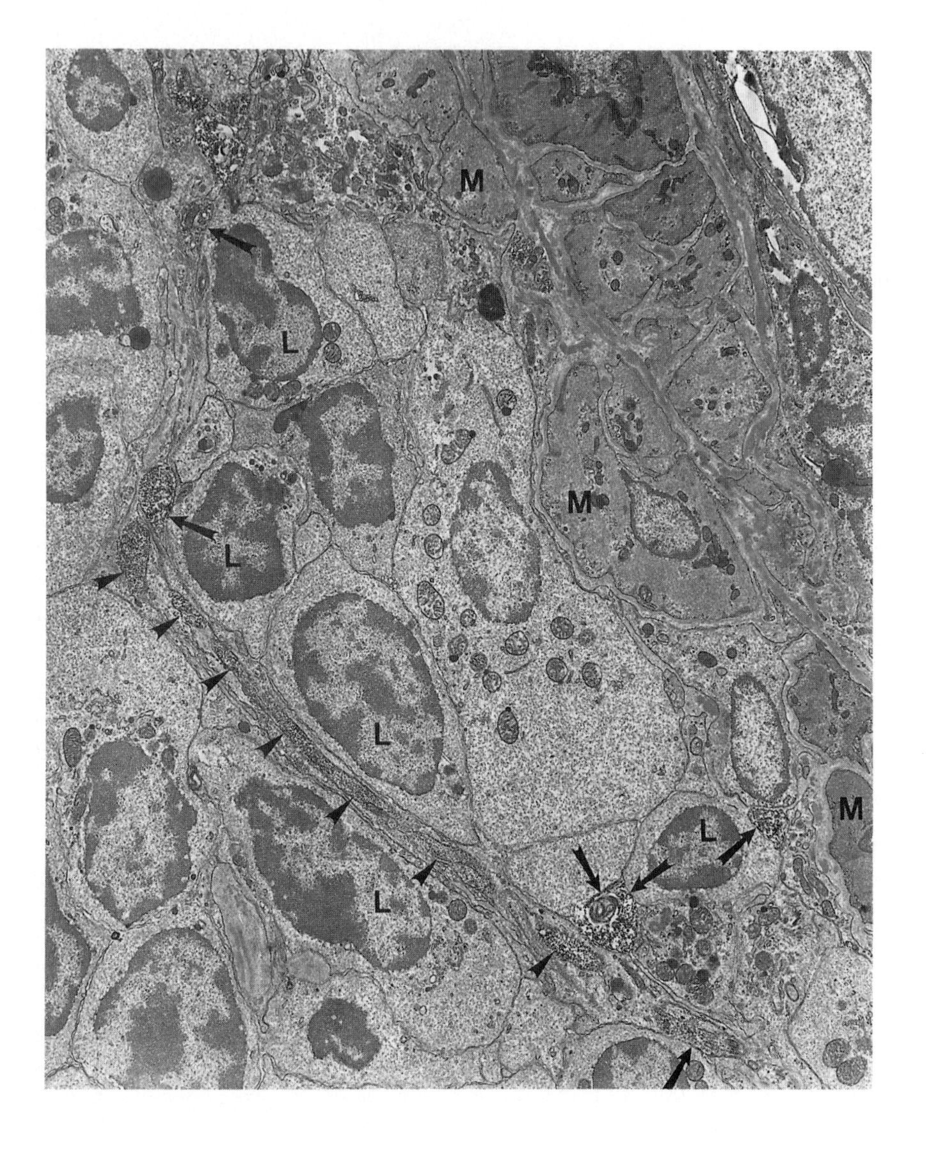

Fig. 28. A bundle of TH-positive nerve terminals is present in the PALS, among lymphocytes, several cell layers deep to the smooth muscle cells (M) of a central arteriole in the white pulp. Some of the TH-positive nerve terminals directly contact (arrows) lymphocytes (L) along this nerve bundle, while other terminals show no such direct contacts (arrowheads). The lymphocyte in the lower right has two TH-positive terminals contacting it. Fischer 344 rat. EM immunohistochemistry. ×8,700.

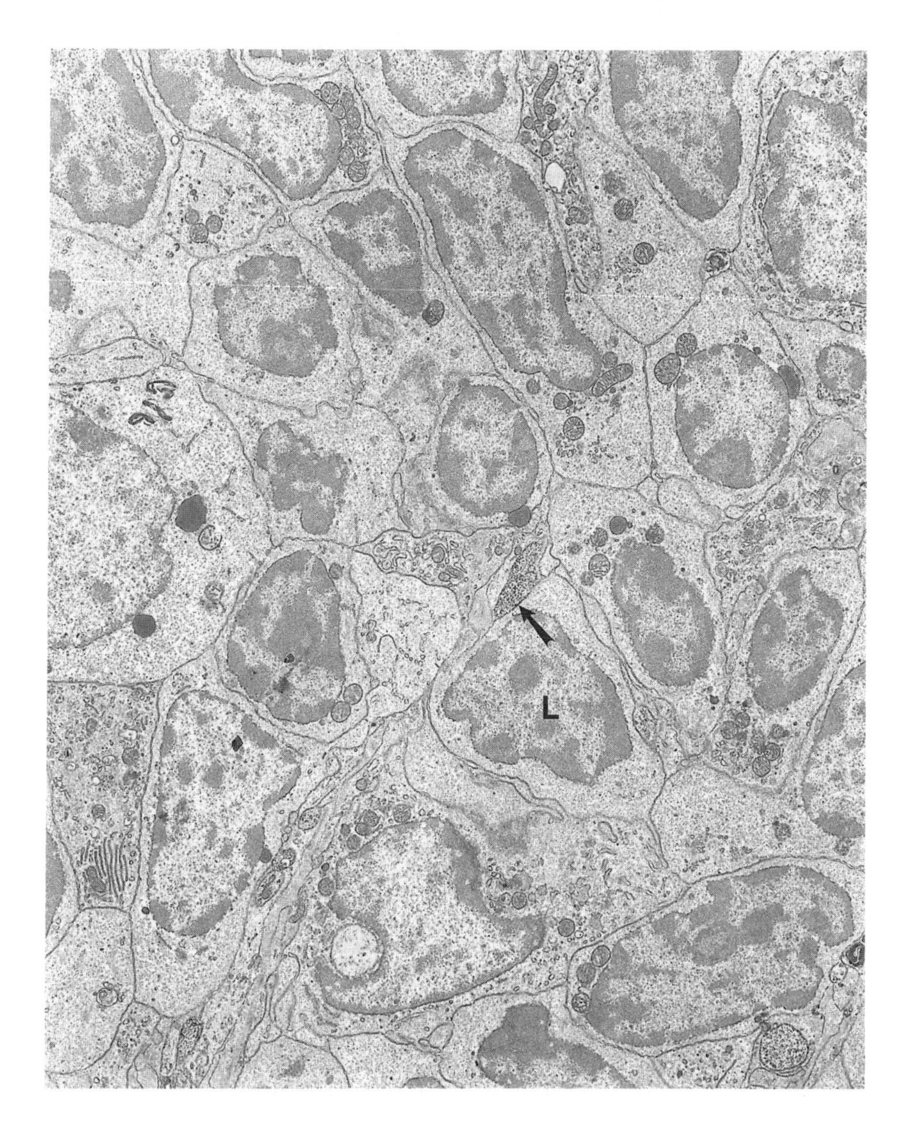

Fig. 29. A TH-positive nerve terminal contacting (arrow) a lymphocyte (L) deep within the PALS of the splenic white pulp. This terminal is part of a nerve bundle that is distant from the vasculature. Fischer 344 rat. EM immunohistochemistry. ×8,700.

ing nerves distribute among the cells of the immune system that bear receptors for the neuropeptide, and are anatomically positioned to provide direct signaling from nerve to lymphoid cell.

Innervation of the Bone Marrow and Thymus

Varicose noradrenergic nerve fibers enter the bone marrow of the rat as arteriolar plexuses, follow those vessels into the marrow, and then further distribute into the parenchyma [36], in agreement with the earlier light and electron microscopic observations of Calvo [17] and Calvo and Forteza-Vila [18]. Nerve terminals end among hematopoietic and lymphopoietic cells of the marrow, and provide the anatomical substrate for delivering neurotransmitter into the extracellular environment of these mobile cells. Of course, for this hypothesis to be substantiated, adrenergic receptors and appropriate second messenger links in the target cells need to be identified. We have observed SP-immunoreactive fibers in some sections of bone marrow, but have not mapped these in detail.

Noradrenergic fibers enter the thymus with the vasculature in both mice and rats as a tangled nerve plexus (fig. 1). Some fibers continue with the vasculature into the cortex of the thymus, while others form a capsular/subcapsular plexus. The innervation of the thymus distributes extensively to the cortex (fig. 5), with less innervation of the medulla [11, 31, 36, 43, 61, 100, 101], in general agreement with other studies [16, 66, 92]. Although the noradrenergic fibers are distributed throughout the cortex, a particularly dense pattern is seen in the outermost cortex, where the immature thymocytes reside, and at the corticomedullary junction (fig. 6). As the thymus involutes with age, this pattern becomes even more dense [11]. The noradrenergic varicosities are not only present along vasculature in the thymic cortex, but also branch abundantly into the parenchyma. The possibility that NE from these varicosities can influence thymocytes is strengthened by the demonstration of beta-adrenergic receptors on developing thymocytes [93] and the ability of catecholamines to influence the responsiveness of the expression of T alloantigens on these cells [93]. Although NE is present in the thymus [101], it is not known to what extent it is available as a paracrine secretion, as a neurotransmitter in direct synaptic-like contacts, or both. It also is possible that other cells may be targets of NE in the thymus, including mast cells, eosinophils, or large granular autofluorescent cells [16, 101].

NPY generally follows the distribution of noradrenergic fibers, with which it is co-localized [13]. SP and CGRP nerve fibers distribute in the rat thymus along the capsule (fig. 7) and intralobular septa (fig. 8), generally free of vascular association, and enter the thymic cortex from the septa [65]. SP fibers distribute among cortical thymocytes and mast cells. These peptidergic fibers also distribute along corticomedullary vasculature, and are sparse in the medulla. VIP fibers distribute mainly along the capsule and intralobular septa and distribute in zones where mast cells are abundant.

Innervation of the Spleen

Numerous studies, some dating back several decades [87, 96], have shown the presence of nerve fibers in the spleen of all species studied (fig. 9–11). Fluorescence histochemical observations have revealed the abundant presence of noradrenergic fibers as a major component of this innervation [21, 22, 36, 46, 49, 61, 85, 86, 100, 101, 103]. Many of these studies focused on the innervation of smooth muscle of the vasculature or the trabecular/capsular system, consistent with physiological evidence that splenic nerve stimulation [90] or application of NE [reviewed by 23] result in splenic capsular/trabecular contraction, and that NE can regulate splenic vascular resistance and blood flow [8, 83]. In addition to these possible smooth muscle targets of noradrenergic innervation, cells of the splenic white pulp also may be targets of these fibers, as suggested by fluorescence histochemical studies [36, 61, 101] and immunocytochemical studies [2, 4, 28, 30, 33–35, 37, 43, 65].

In the rat, noradrenergic fibers that innervate the spleen originate from cell bodies in the superior mesenteric-celiac ganglion [12], form the splenic nerve, and follow the splenic arterial system into the hilar region of the spleen. Upon entering the spleen, the noradrenergic fibers either continue with the vasculature, or contribute to a dense capsular and trabecular system of innervation. This pattern is consistent among species that contain an abundance of trabecular smooth muscle as well as species that contain very little trabecular smooth muscle. Most of the noradrenergic (and tyrosine hydroxylase (TH)-positive) fibers are found in the white pulp, with only occasional ($<1\%$) profiles found in the red pulp. The TH-positive fibers form plexuses around the central artery and its branches in the white pulp; additional fibers radiate into the parenchyma of the PALS, often running in parallel with the vasculature several cells layers away from the smooth muscle (fig. 12, 13). Observations with double-labelled immunocytochemistry for TH and specific markers for cells of the immune system (OX-19 for pan T cells; OX-8 for T suppressor cells; W3/25 for T helper cells) have shown the nerve fibers and terminals directly adjacent to T cells, both helper and suppressor subsets, in the PALS (fig. 14) [2, 4, 28]. Similar dual staining procedures using the ED-3 antibody for antigen-presenting macrophages has shown the close association of TH-positive nerve fibers with these macrophages at the marginal sinus (fig. 15). Additional TH-positive fibers are distributed farther out into the marginal zone, and in a smooth parafollicular location around all margins of the follicles. Only occasionally are nerve fibers present within a follicle. The general lack of association of noradrenergic fibers with B lymphocytes may only be a feature of adult innervation; during ontogeny of the spleen during the first 2 weeks of life, before the formation of follicles, the periarteriolar regions populated by B lymphocytes are innervated abundantly by noradrenergic nerve fibers.

The apparent close association of TH-positive nerve fibers and terminals with T lymphocytes and macrophages at the light microscopic level led us to examine these relationships at the electron microscopic (EM) level [33, 41–45]. Nerve terminals were identified with EM immunocytochemistry by their DAB reaction product, their size and appearance, and their contacts with target cells. The central arterioles were selected as the first region for these observations due to the clear compartmentation and identification of cells. TH-positive terminals are present in the adventitia of the central arterioles, always separated from the smooth muscle cells by at least the basement membrane, and sometimes by additional cell processes. Therefore, these classical 'neuro-effector junctions' with smooth muscle, by necessity, must exert their influence by a paracrine-like diffusion of NE rather than a close synaptic contact. Some of these same TH-positive terminals in the adventitia directly contact lymphocytes facing away from the vascular smooth muscle (fig. 25, 26). These contacts form smooth membrane appositions with approximately 6 nm separation, frequently with slightly thickened membranes on both sides. Some of these terminals form indentations in the lymphocyte surface (fig. 27). These synaptic-like contacts are found not only at the junction of the PALS with the arteriolar smooth muscle, but also are present abundantly within the deeper regions of the PALS and at the marginal sinus (fig. 26, 28, 29). Within the PALS, bundles of TH-positive nerve fibers and terminals travel for long distances forming numerous contacts with lymphocytes throughout their course (fig. 28). Thus, it appears that lymphocytes are not only subject to possible influences by the paracrine-like secretion and diffusion of NE, but may be contacted directly by close junctional appositions.

Earlier studies at the EM level have reported some nerve fibers adjacent to lymphocytes, reticular cells, or erythrocytes [15, 48, 73, 84, 103]; these observations were viewed either with considerable caution [15], or were dismissed as mistaken identification of platelets rather than nerve terminals [53]. The identification of the TH-positive profiles as nerve terminals is clear at the EM level, and is further substantiated by the loss of TH-positive profiles following denervation either by ganglionectomy or administration of 6-hydroxydopamine. These nerve terminals are adjacent to cells of the immune system, and certainly are far more regular and close associations than those associated with the vascular smooth muscle cells. These findings suggest that these nerve fibers may influence lymphocytes and macrophages directly in addition to possible regulation of vascular flow or initiation of capsular/trabecular contraction. This nontraditional form of innervation of lymphoid cells that probably are mobile may represent a major link between the CNS and the immune system.

NPY distribution in the spleen follows the same general pattern as that of the noradrenergic, TH-positive fibers (fig. 16). SP and CGRP nerve fibers enter the spleen with the splenic artery in the hilar region, travel along the venous sinuses and trabecular (fig. 17), and branch into the red pulp (fig. 18) [13, 64]. Additional nerve fibers also were found in the marginal zone and as long linear profiles in the outer PALS of the white pulp (fig. 19). Lundberg et al. [66] reported NPY, SP, and VIP-containing nerve fibers along the splenic vasculature.

Innervation of Lymph Nodes

Noradrenergic TH-positive nerve fibers enter cervical, mesenteric, and popliteal lymph nodes with the vasculature at the hilar region in rats and mice (fig. 20) [2, 38, 100, 101]; this innervation is less extensive than noradrenergic innervation of the spleen. The fibers either continue with the vasculature into the medullary cords, or contribute to a capsular/subcapsular plexus. Within the medullary cords, the fibers run adjacent to the vasculature and freely through cellular regions. These fibers continue past the corticomedullary junction and distribute into the paracortical and cortical regions, where T lymphocytes are found (fig. 21). Additional noradrenergic fibers contribute to this innervation from the subcapsular region. The noradrenergic innervation therefore distributes to the medullary cords, the paracortical and cortical regions, the parafollicular areas, the subcapsular sinus, and the capsule, but generally avoids the follicles themselves.

The patterns of noradrenergic innervation of the lymph nodes and spleen show similarities that may reflect a common role. The innervation distributes to sites of entry of T and B lymphocytes, sites of antigen capture, sites of antigen presentation and T lymphocyte activation, and sites of egress of lymphocytes. The possibility that NE may influence some of these functions is strengthened by functional studies [61] showing a marked diminution of primary and secondary immune responses following denervation of spleen and lymph nodes, and studies showing an influence of catecholamines on egress of activated lymphocytes from the spleen [25, 26] and lymph nodes [74]. It is unlikely that we can identify 'the' role of NE in the spleen or lymph nodes; it is likely that this neurotransmitter can act on numerous cell types in many compartments of secondary lymphoid organs, probably in the presence of differing concentrations of other neurotransmitters and cytokines, to act upon adrenergic receptors found in varying densities on the target cells that may be differentially responsive under varying states of activation. We suggest that at the very least, NE may represent another immunomodulatory substance with considerable functional complexity.

Neuropeptide nerve fibers also are present in lymph nodes. NPY fibers follow the same general course and localization (fig. 22) as noradrenergic nerve fibers, although they may not always be co-localized with NE. VIP fibers were noted along the vasculature in nodal regions of the cortex, with sparse innervation elsewhere [48, 80]. Dynorphin A and CCK nerves were observed in the medulla of guinea pig lymph nodes [59]. SP and CGRP nerve fibers appear to be co-localized and are present in the hilus, beneath the capsule, at the corticomedullary junction, in medullary cords, and in internodal regions (fig. 23) [13]. These nerves are found along the vasculature and among cells of the immune system in the parenchyma.

Gut-Associated Lymphoid Tissue (GALT)

Noradrenergic fibers innervate GALT, both in Peyer's patches and in specialized immune structures such as the rabbit appendix and sacculus rotundus [39, 55]. Noradrenergic fibers arise from ganglion cells in the mesenteric ganglia, and enter the gut at the serosal surface in association with large vascular nerve plexuses. These fibers distribute inside the muscularis interna, travel in association with smooth muscle of the gut and blood vessels, and then turn radially to run in internodular septa between the lymph nodules. The nerve plexuses travel towards the lumen of the gut, course directly through the T-dependent area at the base of the lamina propria (fig. 24), and then enter the lamina propria itself. The nerve fibers arborize abundantly in the lamina propria, forming a dense terminal network in the subepithelial region among plasma cells. Noradrenergic terminals appear to form additional associations with lymphocytes and enterochromaffin cells (some which appear to be serotonin-containing) in the lamina propria. No innervation appeared to be present in the nodules or in the crowns of the domes, where B lymphocytes are abundant. Some acetylcholinesterase (AChE)-positive profiles are present, but are extremely difficult to interpret, since many noradrenergic fibers also are AChE-positive, and most AChE-positive profiles clearly are nonneural. It therefore is highly unlikely that this histochemical method can be used to identify reliably cholinergic nerves, in GALT or in any other lymphoid organ. Such cholinergic innervation, if it exists, will await demonstration with a reliable method such as immunocytochemistry for choline acetyltransferase in peripheral tissue, a technique that has been elusive and unreliable up to this point in the periphery in general, and in lymphoid organs in specific.

SP, CGRP, VIP and somatostatin nerves are present in the lamina propria of Peyer's patches, where most of the immunological effector cells are found [20, 24, 81, 95]. The SP and CGRP fibers appear closely associated with mast cells in the jejunal intestinal mucosa [95]. Ottaway et al. [79] have identified a plexus of VIP nerve fibers coursing with small caliber blood vessels of Peyer's patches.

Criteria for Neurotransmission

In order for NE, or any other putative neurotransmitter, to be accepted as a bona fide neurotransmitter by classical evaluation, it must meet several criteria. The first criterion is the presence and localization of nerve fibers that synthesize the compound in the region where the target cells reside. The sections above document the presence of noradrenergic and peptidergic nerve fibers localized in precise compartments of both primary and secondary lymphoid organs. The second criterion for neurotransmission is release, a criterion for which there is mainly indirect evidence for NE and for which little investigation has been directed for neuropeptides. The third criterion for neurotransmission is the presence of receptors on the target cells, for which an abundance of evidence is available for both adrenergic and peptidergic receptors on lymphocytes, macrophages, and other cells of the immune system [for a detailed review, see 1, 13]. The fourth criterion for neurotransmission is a functional role for the innervation, involving the specific neurotransmitter, that can be revealed by pharmacological or physiological manipulation. In the case of noradrenergic innervation, we have accumulated evidence that NE may inhibit neutrophil oxidative burst activity, inhibit thymocyte proliferation, enhance T-dependent immune responses, enhance cytotoxic T cell activity and delayed hypersensitivity responses, and exert an inhibitory effect (in lymph nodes) on the severity of expression of adjuvant induced arthritis. We have been unable to reliably demonstrate an inhibitory role for NE in T cell proliferation. These functional studies are highly complex, reveal possible multiple effects depending on the timing of denervation or stimulation, and may involve simultaneous actions of several mediators on multiple cell types. Detailed reviews of the extensive actions of catecholamines [7, 29, 34, 60–62, 67–69] and neuropeptides [13, 19, 51, 72, 78, 94] on immune functions are available. An assessment of how well these putative neurotransmitters fulfill the criteria for neurotransmission recently has been provided [27].

It now is evident that extensive neural-immune anatomical connections exist between the nervous and immune systems, with close contacts of nerves with lymphocytes and macrophages. The presence of receptors for catecholamines and neuropeptides on these cells, coupled with functional evidence that these neural signals can modulate immune responses, brings these putative neurotransmitters to the forefront as a class of immunomodulatory molecules that can be investigated for possible benefit of disorders resulting from enhanced or suppressed activity of specific aspects of immune function. It certainly is very clear that extensive bidirectional interactions occur between the nervous and immune systems, and that one system cannot be considered functionally without taking into account the state of activity of the other system.

References

1 Ackerman KD, Bellinger DL, Felten SY, Felten DL: Ontogeny and senescence of noradrenergic innervation of the rodent thymus and spleen; in Ader R, Felten DL,

Cohen C (eds): Psychoneuroimmunology, ed 2. San Diego, Academic Press, 1991, pp 71–126.

2 Ackerman KD, Felten SY, Bellinger DL, Felten DL: Noradrenergic sympathetic innervation of the spleen. III. Development of innervation in the rat spleen. J Neurosci Res 1987;18:49–54, 122–125.

3 Ackerman KD, Felten SY, Bellinger DL, Livnat S, Felten DL: Noradrenergic sympathetic innervation of spleen and lymph nodes in relation to specific cellular compartments. Prog Immunol 1987;6:588–600.

4 Ackerman KD, Felten SY, Dijkstra CD, Livnat S, Felten DC: Parallel development of noradrenergic innervation and cellular compartmentation in the rat spleen. Exp Neurol 1989;103:239–255.

5 Ader R, Cohen N: The influence of conditioning on immune responses; in Ader R, Felten DL, Cohen N (eds): Psychoneuroimmunology, ed 2. San Diego, Academic Press, 1991, pp 611–646.

6 Ader R, Cohen N, Grota LJ: Adrenal involvement in conditioned immunosuppression. Int J Immunopharmacol 1979;1:141–145.

7 Ader R, Felten DL, Cohen N: Interactions between the brain and the immune system. Annu Rev Pharmacol Toxicol 1990;30:561–602.

8 Ayers A, Davies BN, Withrington PG: Responses of the isolated, perfused human spleen to sympathetic nerve stimulation, catecholamines and polypeptides. Br J Pharmacol 1972;44:17–30.

9 Bellinger DL, Ackerman KD, Felten SY, Lorton D, Felten DL: Noradrenergic sympathetic innervation of thymus, spleen and lymph nodes: Aspects of development, aging and plasticity in neural immune interaction; in Hadden JW, Masek K, Nistrio G (eds): Interactions among Central Nervous System, Neuroendocrine and Immune Systems. Rome, Pythagora Press, 1989, pp 35–66.

10 Bellinger DL, Felten SY, Collier TJ, Felten DL: Noradrenergic sympathetic innervation of the spleen. IV. Morphometric analysis in adult and aged F344 rats. J Neurosci Res 1987;18:55–63.

11 Bellinger DL, Felten SY, Felten DL: Maintenance of noradrenergic sympathetic innervation in the involuted thymus of the aged Fischer 344 rat. Brain Behav Immun 1988;2:133–150.

12 Bellinger DL, Felten SY, Lorton D, Felten DL: Origin of noradrenergic innervation of the spleen in rats. Brain Behav Immun 1989;3:291–311.

13 Bellinger DL, Lorton D, Romano TD, Olschowka JA, Felten SY, Felten DL: Neuropeptide innervation of lymphoid organs. Ann NY Acad Sci 1990;594: 17–33.

14 Bernton EW, Bryant HU, Holaday JW: Prolactin and immune function; in Ader R, Felten DL, Cohen N (eds): Psychoneuroimmunology, ed 2. San Diego, Academic Press, 1991, pp 403–428.

15 Blue J, Weiss L: Electron microscopy of the red pulp of the dog spleen including vascular arrangements, periarterial macrophage sheaths (ellipsoids), and the contractile, innervated reticular meshwork. Am J Anat 1981;161:189–218.

16 Bulloch K, Pomerantz W: Autonomic nervous system innervation of thymic-related lymphoid tissue in wild-type and nude mice. J Comp Neurol 1984;228:57–68.

17 Calvo W: The innervation of the bone marrow in laboratory animals. Am J Anat 1968;123:315–328.

18 Calvo W, Forteza-Vila J: On the development of bone marrow innervation in

newborn rats as studied with silver impregnation and electron microscopy. Am J Anat 1969;126:355–359.
19 Carr DJJ, Blalock JE: Neuropeptide hormones and receptors common to the immune and neuroendocrine systems: Bidirectional pathway of intersystem communication; in Ader R, Felten DL, Cohen N (eds): Psychoneuroimmunology, ed 2. San Diego, Academic Press, 1991, pp 573–588.
20 Cooke JH: Neurobiology of the intestinal mucosa. Gastroenterology 1986;90:1058–1081.
21 Dahlström AB, Häggendal J, Hökfelt T: The noradrenaline content of the varicosities of sympathetic adrenergic nerve terminals in the rat. Acta Physiol Scand 1966; 67:289.
22 Dahlström AB, Zetterström BEM: Noradrenaline stores in nerve terminals of the spleen: Changes during hemorrhagic shock. Science 1965;147:1583–1585.
23 Davies BN, Withrington PG: The actions of drugs on the smooth muscle of the capsule and blood vessels of the spleen. Pharmacol Rev 1973;25:373–412.
24 Ekblad E, Winther C, Ekman R, Hakanson R, Sundler F: Projections of peptide-containing neurons in rat small intestine. Neuroscience 1987;20:169–188.
25 Ernström U, Sandberg G: Effects of alpha- and beta-receptor stimulation on the release of lymphocytes and granulocytes from the spleen. Scand J Haematol 1973;11:275–286.
26 Ernström U, Soder O: Influence of adrenaline on the dissemination of antibody-producing cells from the spleen. Clin Exp Immunol 1975;21:131–140.
27 Felten DL: Neurotransmitter signaling of cells of the immune system: important progress, major gaps. Brain Behav Immun 1991;5:2–8.
28 Felten DL, Ackerman KD, Wiegand SJ, Felten SY: Noradrenergic sympathetic innervation of the spleen. I. Nerve fibers associate with lymphocytes and macrophages in specific compartments of the splenic white pulp. J Neurosci Res 1987;18:28–36, 118–121.
29 Felten DL, Bellinger DL, Felten SY: Autonomic signaling of the immune system: implication for neural-immune interactions in aging; in Hendrie H, Mendelson LG, Readhead C (eds): Brain Aging: Molecular Biology, the Aging Process and Neurodegenerative Disease. Toronto, Hogrefe & Huber, 1990, pp 239–258.
30 Felten DL, Felten SY: Sympathetic noradrenergic innervation of immune organs. Brain Behav Immun 1988;2:293–300.
31 Felten DL, Felten SY: Innervation of the thymus; in Kendall M (ed): Thymus Update 2. London, Harwood Academic, 1989, pp 73–88.
32 Felten DL, Felten SY: Brain-immune signaling: substrate for reciprocal immunological signaling of the nervous system, in Frederickson RCA, McGaugh JL, Felten DL (eds): Peripheral Signaling of the Brain: Role in Neural-Immune Interactions, Learning and Memory. Toronto, Hogrefe & Huber, 1991, pp 3–17.
33 Felten DL, Felten SY, Ackerman KD, Bellinger DL, Madden KS, Carlson SL, Livnat S: Peripheral innervation of lymphoid tissue; in Freier S (ed): The Neuroendocrine-Immune Network. Boca Raton, CRC Press, 1990, pp 9–18.
34 Felten DL, Felten SY, Bellinger DL, Carlson SL, Ackerman KD, Madden KS, Olschowka JA, Livnat S: Noradrenergic sympathetic neural interactions with the immune system: Structure and function. Immunol Rev 1987;100:225–260.
35 Felten DL, Felten SY, Carlson SL, Bellinger DL, Ackerman KD, Romano TA, Livnat S: Development, aging, and plasticity of noradrenergic sympathetic innerva-

tion of secondary lymphoid organs: implications for neural-immune interactions; in Dahlström A, Belmaker RH, Sandler M (eds): Progress in Catecholamine Research. A. Basic Aspects and Peripheral Mechanisms. New York, Liss, 1988, pp 517–524.
36 Felten DL, Felten SY, Carlson SL, Olschowka JA, Livnat S: Noradrenergic and peptidergic innervation of lymphoid tissue. J Immunol 1985;135:755s–765s.
37 Felten DL, Felten SY, Madden KS, Ackerman KD, Bellinger DL: Development, maturation and senescence of sympathetic innervation of secondary immune organs; in Schreibman MP, Scanes CG (eds): Development, Maturation and Senescence of the Neuroendocrine System. San Diego, Academic Press, 1989, pp 381–396.
38 Felten DL, Livnat S, Felten SY, Carlson SL, Bellinger DL, Yeh P: Sympathetic innervation of lymph nodes in mice. Brain Res Bull 1984;13:693–699.
39 Felten DL, Overhage JM, Felten SY, Schmedtje JF: Noradrenergic sympathetic innervation of lymphoid tissue in the rabbit appendix: further evidence for a link between the nervous and immune systems. Brain Res Bull 1981;7:595–612.
40 Felten SY, Bellinger DL, Collier TJ, Coleman PD, Felten DL: Decreased sympathetic innervation of spleen in aged Fischer 344 rats. Neurobiol Aging 1987;8:159–165.
41 Felten SY, Felten DL: Are lymphocytes targets of noradrenergic innervation? In Weiner H, Helhammer D, Murison R, Florin I (eds): Frontiers of Stress Research. Toronto, Huber, 1989, pp 56–71.
42 Felten SY, Felten DL: Sympathetic noradrenergic neural contacts with lymphocytes and macrophages in the splenic white pulp of the rat: Site of possible bidirectional communication and local regulation between the nervous and immune system; in Bunney WE Jr, Hippius H, Laakman G, Schmauss M (eds): Neuropsychopharmacology. Berlin, Springer, 1990, pp 442–456.
43 Felten SY, Felten DL: The innervation of lymphoid tissue; in Ader R, Felten DL, Cohen N (eds): Psychoneuroimmunology, ed 2. San Diego, Academic Press, 1991, pp 27–69.
44 Felten SY, Olschowka J: Noradrenergic sympathetic innervation of the spleen. II. Tyrosine hydroxylase (TH)-positive nerve terminals form synaptic-like contacts on lymphocytes in the splenic white pulp. J Neurosci Res 1987;18:37–44.
45 Felten SY, Olschowka J, Ackerman KD, Felten DL: Catecholaminergic innervation of the spleen: Are lymphocytes targets of noradrenergic nerves? In Dahlstrom A, Belmaker RH, Sandler M (eds): Progress in Catecholamine Research. A. Basic Aspects and Peripheral Mechanisms. New York, Liss, 1988, pp 539–546.
46 Fillenz M: The innervation of the cat spleen. Proc R Soc Lond 1970;174:459–468.
47 Fink T, Weihe E: Multiple neuropeptides in nerves supplying mammalian lymph nodes: messenger candidates for sensory and autonomic neuroimmunomodulation. Neurosci Lett 1988;90:39–44.
48 Galindo B, Imaeda T: Electron microscopic study of the white pulp of the mouse spleen. Anat Rec 1962;143:399–415.
49 Gillespie JS, Kirpekar SM: The localization of endogenous and infused noradrenaline in the spleen. J Physiol 1965;179:46P.
50 Giron LT, Crutcher KA, Davis JN: Lymph nodes – a possible site for sympathetic neuronal regulation of immune responses. Ann Neurol 1980;8:520–522.
51 Goetzl EJ, Turck CW, Sreedharan SP: Production and recognition of neuropeptides by cells of the immune system; in Ader R, Felten DL, Cohen N (eds): Psychoneuroimmunology, ed 2. San Diego, Academic Press, 1991, pp 263–282.
52 Heijnen CJ, Kavelaars A, Ballieux RE: Corticotropin-releasing hormone and pro-

opiomelanocortin-derived peptides in the modulation of immune function; in Ader R, Felten DL, Cohen N (eds): Psychoneuroimmunology, ed 2. San Diego, Academic Press, 1991, pp 429–446.
53 Heusermann U, Stutte HJ: Electron microscopic studies of the innervation of the human spleen. Cell Tissue Res 1977;184:225–236.
54 Irwin M, Vale W, Britton KT: Central corticotropin-releasing factor suppresses natural killer cytotoxicity. Brain Behav Immun 1987;1:81–87.
55 Jesseph JM, Felten DL: Noradrenergic innervation of the gut-associated lymphoid tissues (GALT) in the rabbit. Anat Rec 1984;208:81A.
56 Keller SE, Weiss JM, Schleiffer SJ, Miller NE, Stein M: Suppression of immunity in adrenalectomized rats. Science 1983;221:1301–1304.
57 Kelley KW: Growth hormone in immunobiology; in Ader R, Felten DL, Cohen N (eds): Psychoneuroimmunology, ed 2. San Diego, Academic Press, 1991, pp 377–402.
58 Kendall MD, Al-Shawaf AA: Innervation of the rat thymus gland. Brain Behav Immun 1991;5:9–28.
59 Kurkowski R, Kummer W, Heym C: Substance P-immunoreactive nerve fibers in tracheobronchial lymph nodes of the guinea pig: origin, ultrastructure and coexistence with other peptides. Peptides 1990;11:13–20.
60 Livnat S, Eisen J, Felten DL, Felten SY, Irwin J, Madden KS, Sundaresan PJ: Behavioral and sympathetic neural modulation of immune function; in Dahlström A, Belmaker RM, Sandler M (eds): Progress in Catecholamine Research. A. Basic Aspects and Peripheral Mechanisms. New York, Liss, 1988, pp 539–546.
61 Livnat S, Felten SY, Carlson SL, Bellinger DL, Felten DL: Involvement of peripheral and central catecholamine systems in neural-immune interactions. J Neuroimmunol 1985;10:5–30.
62 Livnat S, Madden KS, Felten DL, Felten SY: Regulation of the immune system by sympathetic neural mechanisms. Prog Neuropsychopharmacol Biol Psychiatry 1987;11:145–152.
63 Lorton D, Bellinger DL, Felten SY, Felten DL: Substance P innervation of the rat thymus. Peptides 1990;11:1269–1275.
64 Lorton D, Bellinger DL, Felten SY, Felten DL: Substance P innervation of spleen in rats: nerve fibers associated with lymphocytes and macrophages in specific compartments of the spleen. Brain Behav Immun 1991;5:29–40.
65 Lorton D, Hewitt D, Bellinger DL, Felten SY, Felten DL: Noradrenergic reinnervation of the rat spleen following chemical sympathectomy with 6-hydroxydopamine: Pattern and time course of reinnervation. Brain Behav Immun 1990;4:198–222.
66 Lundberg JM, Anggard A, Pernow J, Hökfelt T: Neuropeptide Y-, substance P- and VIP-immunoreactive nerves in cat spleen in relation to autonomic vascular and volume control. Cell Tissue Res 1985;239:9–18.
67 Madden KS, Ackerman KD, Livnat S, Felten SY, Felten DL: Patterns of noradrenergic innervation of lymphoid organs and immunological consequences of denervation; in Goetzl E (ed): Physiology and Diseases. New York, Liss, 1989.
68 Madden KS, Felten SY, Felten DL, Livnat S: Sympathetic neural modulation of the immune system. I. Depression of T cell immunity in vivo and in vitro following chemical sympathectomy. Brain Behav Immun 1989;3:72–89.
69 Madden KS, Livnat S: Catecholamine action and immunologic reactivity; in Ader R,

Felten DL, Cohen N (eds): Psychoneuroimmunology, ed 2. San Diego, Academic Press, 1991, pp 283–310.
70 Maestroni GJM, Conti A: Role of the pineal neurohormone melatonin in the psychoneuroendocrine-immune network; in Ader R, Felten DL, Cohen N (eds): Psychoneuroimmunology, ed 2. San Diego, Academic Press, 1991, pp 495–514.
71 McCruden AB, Stimson WH: Sex hormones and immune function; in Ader R, Felten DL, Cohen N (eds): Psychoneuroimmunology, ed 2. San Diego, Academic Press, 1991, pp 475–494.
72 McGillis JP, Mitsuchashi M, Payan DG: Immunologic properties of substance P; in Ader R, Felten DL, Cohen N (eds): Psychoneuroimmunology, ed 2. San Diego, Academic Press, 1991, pp 209–224.
73 Moore RD, Mumaw VR, Schoenberg MC: The structure of the spleen and its functional implications. Exp Mol Pathol 1964;3:31–50.
74 Moore TC: Modification of lymphocyte traffic by vasoactive neurotransmitter substances. Immunology 1984;52:511–518.
75 Müller S, Weihe E: Interrelation of peptidergic innervation with mast cells and ED1-positive cells in rat thymus. Brain Behav Immun 1991;5:55–72.
76 Munck A, Guyre PM: Glucocorticoids and immune function; in Ader R, Felten DL, Cohen N (eds): Psychoneuroimmunology, ed 2. San Diego, Academic Press, 1991, pp 447–474.
77 Nohr D, Weihe E: The neuroimmune link in the bronchus-associated lymphoid tissue (BALT) of cat and rat: peptides and neural markers. Brain Behav Immun 1991;5:84–101.
78 Ottaway CA: Vasoactive intestinal peptide and immune function; in Ader R, Felten DL, Cohen N (eds): Psychoneuroimmunology, ed 2. San Diego, Academic Press, 1991, pp 225–262.
79 Ottaway CA, Lewis DL, Asa SL: Vasoactive intestinal peptide-containing nerves in Peyer's patches. Brain Behav Immun 1987;1:148–158.
80 Popper P, Mantyh CR, Vigna SA, Magioos JE, Mantyh PW: The localization of sensory nerve fibers and receptor binding sites for sensory neuropeptides in canine mesenteric lymph nodes. Peptides 1988;9:257–267.
81 Probert L, de Mey J, Polak JM: Distinct subpopulations of enteric p-type neurones contain substance P and vasoactive intestinal polypeptide. Nature 1981;294:470–471.
82 Rabin BS, Cunnick JE, Lysle DT: Stress-induced alteration of immune function. Prog NeuroEndocrinImmunol 1990;3:116–124.
83 Reilly FD: Innervation and vascular pharmacodynamics of the mammalian spleen. Experientia 1985;41:187–192.
84 Reilly Fe, McCuskey PA, Miller ML, McCuskey RS, Meineke HA: Innervation of the periarteriolar lymphatic sheath of the spleen. Tissue Cell 1979;11:121–126.
85 Reilly FD, McCuskey RS: Studies of the hemopoietic microenvironment. VII. Neural mechanisms in splenic microvascular regulation in mice. Microvasc Res 1977;14:293–302.
86 Reilly FD, McCuskey RS, Meineke HA: Studies of the hemopoietic microenvironment. VIII. Adrenergic and cholinergic innervation of the murine spleen. Anat Rec 1976;185:109–118.
87 Riegele L: Über die mikroskopische Innervation der Milz. Z Zellforsch Mikrosk Anat 1929;9:511–533.

88 Romano TA, Felten SY, Felten DL, Olschowka JA: Neuropeptide-Y innervation of that rat spleen: another potential immunomodulatory neuropeptide. Brain Behav Immun 1991;5:116–131.
89 Sanders SK, Becker KJ, Cierpial MA, Carpenter DM, Rankin LA, Fleener SL, Ritchie JC, Simson PE, Weiss JM: Intracerebroventricular infusion of interleukin-1 rapidly decreases peripheral cellular immune responses. 1989;86:6398–6402.
90 Schiff JM: Leçons sur la physiologie de la digestion. Tome 2, Leçon 1867;35:416.
91 Schultzberg M, Svenson SB, Unden A, Bartfai T: Interleukin-1-like immunoreactivity in peripheral tissues. J Neurosci Res 1987;18:184–189.
92 Sergeeva VE: Histotopography of catecholamines in the mammalia thymus. Bull Exp Biol Med 1974;77:456–458.
93 Singh U, Owen JJT: Studies on the maturation of thymus stem cells – The effects of catecholamines, histamine, and peptide hormones on the expression of T alloantigens. Eur J Immunol 1976;6:59–62.
94 Stead RH, Tomioka M, Pezzati P, Marshall J, Croitoru K, Perdue M, Stanisz A, Bienenstock J: Interaction of the mucosal immune and peripheral nervous systems; in Ader R, Felten DL, Cohen N (eds): Psychoneuroimmunology, ed 2. San Diego, Academic Press, 1991, pp 177–208.
95 Stead RH, Tomioka M, Quinonez G, Simon G, Felten SY, Bienenstock J: Intestinal mucosal mast cells in normal and nematode-infected rat intestines are in intimate contact with peptidergic nerves. Proc Natl Acad Sci USA 1987;84:2975–2979.
96 Utterback RA: The innervation of the spleen. J Comp Neurol 1944;81:55–68.
97 Walcott B, McLean JR: Catecholamine-containing neurons and lymphoid cells in a lacrimal gland of the pigeon. Brain Res 1985;328:129–137.
98 Weihe E, Krekel J: The neuroimmune connection in human tonsils. Brain Behav Immun 1991;5:41–54.
99 Weihe E, Müller S, Fink T, Zentel HJ: Tachykinins, calcitonin gene-related peptide and neuropeptide Y in nerves of the mammalian thymus: interactions with mast cells in autonomic and sensory neuroimmunomodulation. Neurosci Lett 1989;100:77–82.
100 Williams JM, Felten DL: Sympathetic innervation of murine thymus and spleen: A comparative histofluorescence study. Anat Rec 1981;199:531–542.
101 Williams JM, Peterson RG, Shea PA, Schmedtje JF, Bauer DC, Felten DL: Sympathetic innervation of murine thymus and spleen: Evidence for a functional link between the nervous and immune systems. Brain Res Bull 1981;6:83–94.
102 Zentel HJ, Weihe E: The neuro-B cell link of peptidergic innervation in the bursa fabricii. Brain Behav Immun 1991;5:132–147.
103 Zetterström BEM, Hökfelt T, Norberg K-A, Olsson P: Possibilities of a direct adrenergic influence on blood elements in the dog spleen. Acta Chir Scand 1973;139: 117–122.

Suzanne Y. Felten, PhD, Department of Neurobiology and Anatomy,
Box 603, University of Rochester School of Medicine,
601 Elmwood Avenue, Rochester, NY 14642 (USA)

Blalock JE (ed): Neuroimmunoendocrinology, 2nd rev ed.
Chem Immunol. Basel, Karger, 1992, vol 52, pp 49–83

Neuroendocrine Peptide Hormone Regulation of Immunity[1]

Howard M. Johnson, Myron O. Downs, Carol H. Pontzer

Department of Microbiology and Cell Science,
The University of Florida, Gainesville, Fla., USA

Introduction

It is now well established that so-called 'non-neuroendocrine' peripheral tissues such as the gut, pancreas, reproductive organs, lymph nodes, thymus and spleen can produce a variety of hormones or hormone-like activities. An intriguing aspect of hormone production by cells and tissues of the immune system and by closely associated nonimmune tissues such as the gut is their possible regulation of immune function. It is of particular interest that studies carried out over the last few years with lymphocytes stimulated by specific antigens, microbial superantigens, and mitogens have shown that cells of the immune system can produce several polypeptide hormones that are known to be produced also by the pituitary [1]. Perhaps the most notable observation was the demonstration that lymphocyte-derived corticotropin (ACTH) is the same as pituitary-derived ACTH as ascertained by amino acid sequencing [2].

Data presented in this review support the view that neuroendocrine polypeptide hormones can regulate several important immune functions. These include antibody production, lymphocyte cytotoxicity, lymphokine production, hypersensitivity, and macrophage and neutrophil activation and chemotaxis. We will focus particularly on immunoregulation by: neuropeptides derived from the polyprotein proopiomelanocortin (POMC), enkephalins, the posterior pituitary hormones vasopressin and oxytocin, vasoactive intestinal peptide, substance P, the gonadotropins, growth hormone, prolactin, somatostatin, and thyrotropin. The data provide compelling evidence for

[1]Supported in part by grants from the National Institutes of Health. Florida Agricultural Experiment Station, Journal Series No. R-01892.

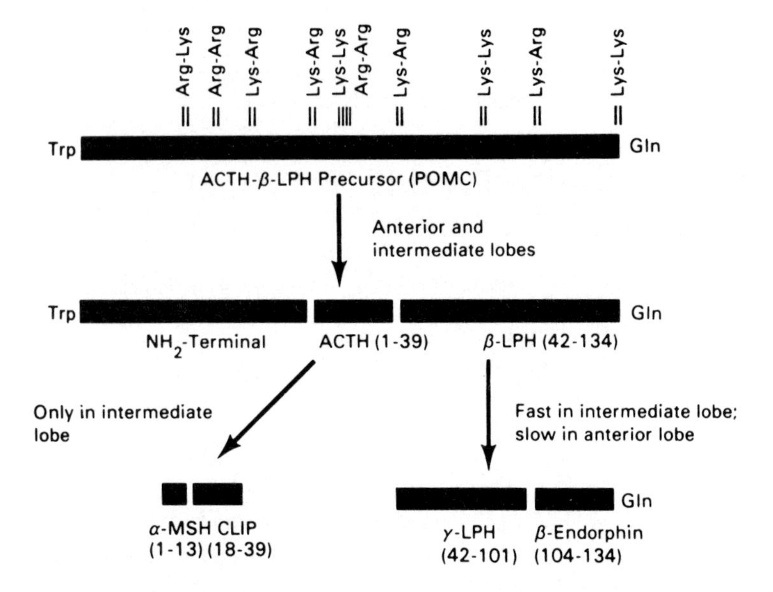

Fig. 1. Pituitary processing of the polyprotein precursor POMC to ACTH, β-endorphin, and other active peptides.

a regulatory circuit between the immune and neuroendocrine systems whereby neuropeptides have a profound effect on positive and negative regulation of immune function at concentrations similar to those required for mediation of their classical functions.

POMC and Proenkephalin Products

Both ACTH and β-endorphin are classically produced in the pituitary gland from the polyprotein precursor POMC (fig. 1) [3]. ACTH is a single-chain polypeptide comprised of 39 amino acids that regulates the stress response in individuals through release of adrenal steroids. An investigation of the possible immunoregulatory role of neuropeptides that are derived from POMC was prompted by the observation that ACTH and endorphin-like activities were produced by lymphocytes that were infected by Newcastle disease virus [4, 5]. ACTH, either synthetic or natural, was a potent inhibitor of antibody production by lymphocytes [6]. This was illustrated in an in vitro plaque-forming cell (PFC) assay [7], where mouse spleen cells readily produced specific antibodies when stimulated with protein antigens such as sheep red blood cells (SRBC) or with carbohydrate complexes such as dinitrophenyl-Ficoll (DNP-Ficoll). ACTH preferentially suppressed the antibody response to protein antigens as compared to carbohydrate-type antigens. Although B lymphocytes are the source of antibodies, helper T lymphocytes are also required for an optimal response to protein antigens like SRBC, while antibody production to DNP-Ficoll is less T cell-dependent [8]. Thus, ACTH may suppress the antibody response in part by blocking helper T cell signals. ACTH also acts at an early stage in the antibody response, where cell-cell interactions such as T cell-B cell collaborations are most likely to occur.

Consistent with ACTH effects on T and B cell function is the demonstration of ACTH receptors on both cell types [9]. The ACTH agonist (^{125}I-Tyr^{23})Phe^{2}-Nle^{4}-ACTH was used to identify two types of binding sites on both normal rat T and B lymphocytes. The lymphocyte receptor displayed characteristics similar to the receptor on adrenal cells. One site has an approximate $K_d = 9 \times 10^{-11}$ *M* with 3,200 sites/cell, and the other had a $K_d = 4 \times 10^{-9}$ *M* with 38,000 sites/cell. The number of high affinity sites could be substantially increased by lipopolysaccharide (LPS) or concanavalin A (Con A) stimulation of B or T cells, respectively. Modulation of receptor number upon cell stimulation suggests that the ACTH receptor may be important in immunoregulation.

The structural basis of ACTH suppression of the antibody response can be ascertained by comparing the ACTH cleavage peptide α-melanocyte-stimulating hormone (α-MSH; $ACTH_{1-13}$ acetylated and amidated) and corticotropin-like intermediate peptide (CLIP; $ACTH_{18-39}$) with $ACTH_{1-39}$ for suppression of the anti-SRBC response (fig. 1). Neither α-MSH nor CLIP could suppress the antibody response [6]. $ACTH_{1-24}$, like $ACTH_{1-39}$, has full steroidogenic activity yet had no effect on antibody production, which suggests a dissociation of the immunoregulatory and steroidogenic properties of $ACTH_{1-39}$.

In contrast to the inhibitory effects of ACTH on antibody production in the murine spleen cell system, others [10] have shown that ACTH enhances the growth and differentiation of enriched cultures of human tonsillar B cells. ACTH at nanomolar concentrations enhanced the proliferation of activated B cells by two- or threefold in the presence of a B cell growth factor (interleukin-4?) or recombinant interleukin-2 (IL-2). Furthermore, ACTH enhanced IL-2 and B cell differentiation factor-induced IgM and IgG immunoglobulin secretion by activated B cells at concentrations similar to those for enhancement of proliferation. ACTH did not affect B cell function in the absence of the interleukins. Additionally, ACTH has been shown to inhibit the in vitro antibody response of primed human peripheral lymphocytes to tetanus toxoid at relatively low concentrations (10^{-13} and 10^{-17} *M*), while enhancing the response at higher concentrations [11]. ACTH as well as β-endorphin also inhibited the secretory immunoglobulin response to a polyclonal mitogen in the mouse system [12]. Thus, it appears that ACTH may either inhibit or enhance B cell function, probably depending on factors such as lymphokines and on the presence of accessory T cells.

The production of the immunoregulatory lymphokine γ interferon (IFNγ) in a mouse spleen cell system is also suppressed by ACTH in a manner similar to suppression of the antibody response [13]. IFNγ production in humans or mice involves macrophages [14] and helper T cells, in addition to the IFNγ-producing cells [15, 16]. ACTH appears to inhibit IFNγ production by interfering with helper cell function, since the blockage can be reversed by factors that restore helper cell competence [13]. The structural requirements for ACTH suppression of IFNγ production are the same as for the suppression of antibody production, with α-MSH, CLIP, and $ACTH_{1-24}$ having no regulatory effect [13].

The endogenous opiates β-, γ-, and α-endorphin are also contained in POMC, and are composed of amino acids 61–91, 61–77, and 61–76, respectively, of β-lipotropin (β-LPH) (fig. 1). The production of ACTH by lymphocytes is associated with production of endorphin-like activities, which suggests a role for a POMC-like precursor in the formation of these lymphocyte-derived hormones [5]. Direct evidence for receptors on mouse spleen cell membranes for endogenous opiates was established in binding studies with ^{3}H[Met]-enkephalin [6]. At least one type of binding site for [Met]-enkephalin was present with a $K_d = 5.9 \times 10^{-10}$ *M*. Indirect evidence for [Met]-enkephalin receptors on human lymphocytes was shown by enhancement of rosette formation by naloxone [17]. Naloxone is an antagonist of α-endorphin and other opiates in that it competes for binding to specific receptors in the brain. A nonopiate receptor for β-endorphin on human lymphocytes has also been reported where naloxone does not block binding [18]. Thus, lymphocytes may contain both opiate and nonopiate receptors for endogenous opiates such as β-endorphin.

The endorphins, β, γ and α, were examined for possible regulation of the in vitro antibody response to SRBC in the murine system [6]. α-Endorphin was a potent inhibitor of the anti-SRBC PFC response at concentrations as low as 5×10^{-8} *M* (table 1). β- and γ-endorphin were minimally inhibitory, in spite of their structural similarity to α-endorphin. The α-endorphin amino

Table 1. Effect of endorphins and enkephalins on the in vitro anti-SRBC PFC response

Hormone	Concentration, μ*M*	PFC/culture ± SD	% suppression
α-Endorphin	0.5	637 ± 246	92
	0.05	2,720 ± 302	60
	0.005	6,333 ± 652	7
β-Endorphin	6	5,320 ± 567	22
γ-Endorphin	5	5,133 ± 580	25
[Leu]-enkephalin	2	2,460 ± 692	64
	0.2	2,520 ± 240	63
	0.02	5,453 ± 670	20
[Met]-enkephalin	2	1,880 ± 250	73
	0.2	4,040 ± 454	41
Control	–	6,840 ± 697	–

C57BL/6 female mice spleen cells (1.5×10^{-7} in 1 ml) were used for the PFC response. Hormones and SRBC were added to cultures on day 0 and PFC responses were determined on day 5.

acid sequence is contained within the amino acid sequences of both β- and γ-endorphin, which suggests that α-endorphin suppression of antibody production is controlled by a stringent signal at the level of ligand-receptor interaction. As shown in table 2, naloxone was able to block α-endorphin-induced suppression of the PFC response. This suggests that α-endorphin suppresses the mouse anti-SRBC PFC response by binding to receptors on the spleen cells that are similar to those of opiate receptors found in the brain.

Although β-endorphin is a poor suppressor of the PFC response relative to α-endorphin, it did compete with α-endorphin for an opiate-like receptor and thus blocked the suppression of the antibody response (table 2). The naloxone and β-endorphin competition data indicate the following: (a) endorphin receptors similar to those in the brain are present on spleen cells and probably lymphocytes in general; (b) α- and β-endorphin probably bind to the same receptors; (c) binding to the endorphin receptor does not necessarily activate immunosuppressive events as illustrated by the blocking of α-endorphin effects by the opiate antagonist naloxone and by β-endorphin. In fact, β-endorphin has been shown to enhance T cell proliferation and IL-2 production [19].

Table 2. Blockage of α-endorphin suppression of the in vitro anti-SRBC PFC response by naloxone and β-endorphin

α-Endorphin μ*M*	Naloxone μ*M*	β-Endorphin μ*M*	PFC/culture	% suppression	p
Experiment 1					
0.5	0	0	853 ± 234	85	<0.01
0.5	3	0	4,567 ± 1,055	18	NS
0	3	0	4,310 ± 1,259	22	
0	0	0	5,540 ± 658	–	
Experiment 2					<0.05
5	0	0	3,220 ± 707	76	NS
5	0	12	7,360 ± 905	46	
0	0	12	6,140 ± 877	55	
0	0	0	13,680 ± 1,018	–	

C57BL/6 female mice spleen cells (1.5×10^7 in 1 ml) were used for the PFC response. Hormones and SRBC were added to cultures on day 0 and PFC responses were determined by day 5. p Values are based on comparisons with the naloxone (n = 3) and β-endorphin (n = 2) controls, respectively. NS = Not significant.

In addition to the murine system, α-endorphin can also suppress the primary in vitro antibody response of human blood lymphocytes to the T cell-dependent antigen ovalbumin [20]. α-Endorphin inhibited antibody production to ovalbumin by blocking both T and B cell function. The suppression probably occurred via binding to the opiate receptor on the lymphocytes, since α-endorphin lacking the N-terminal amino acid tyrosine was not inhibitory. An intact amino terminus is required for α-endorphin binding to its opiate receptors [21].

[Leu]- and [Met]-enkephalin are also endogenous opiates, but have their own polyprotein precursors [22]. Interestingly, [Met]-enkephalin is contained within the N-terminal sequence of all three endorphins (residues 61-65). Both [Leu]- and [Met]-enkephalin were intermediate in their antibody suppressive properties in the mouse spleen cell system, more suppressive than β-endorphin but less suppressive than α-endorphin (table 1) [6]. This suggests that there are stringent steric requirements for endorphin and enkephalin interaction with receptors on the cell membrane in a manner suitable for induction of immunosuppression, with the amino terminal end of the endorphins playing an important role. Others have similarly found β-endorphin and [Met]-enkephalin to suppress the in vitro antibody response [23–25].

Of interest is that rats subjected to opiate-type stress have been reported to show suppressed splenic natural killer (NK) cell activity and lymphocyte proliferation in vitro [26, 27]. This suppression was blocked by the opiate antagonist naltrexone, suggesting that endogenous opiates mediate some forms of stress-induced suppression of the immune response. But this observation was not correlated with similar in vivo reductions in immune function. While elevated circulating corticosterone was observed, it was of insufficient duration to affect the immune response [27]. Other interesting aspects of the immunoregulatory function of POMC-derived peptides are the observations that the opiate peptides α-endorphin, β-endorphin and/or [Met]-enkephalin can: (a) enhance the natural cytotoxicity of lymphocytes and macrophages toward tumor cells [28–33]; (b) enhance or inhibit T cell mitogenesis; the enhancement may be due to induction of IL-2 [19, 34–37]; (c) enhance T cell rosetting [38]; (d) stimulate human peripheral blood mononuclear cell chemotaxis [39, 40]; (e) inhibit T cell chemotactic factor production [41]; (f) inhibit IFNγ production by cultured human peripheral blood mononuclear cells [42], and (g) inhibit major histocompatibility (MHC) class II antigen expression [43]. Inhibition of MHC class II antigen expression is a possible mechanism for the suppressive effects of these hormones.

Arginine Vasopressin

Arginine vasopressin (AVP) is a neurohypophyseal (posterior pituitary) nonapeptide that is classically produced in the magnocellular neurons of the supraoptic and paraventricular nuclei of the hypothalamus. It is transported to the posterior pituitary where it is released into the circulation [44]. Its effects include: (a) antidiuretic activity on specific receptors in the kidney [45]; (b) vasopressor activity which, in concert with other peptide hormones, regulates systemic arterial pressure [46]; (c) modulation of the stress response by direct stimulation of release of ACTH and by enhancement of ACTH release by corticotropin releasing factor [47]; (d) enhancement of learning and memory, functioning as a neurotransmitter [48]. Recently, a number of peripheral tissues and organs have been shown to also produce AVP or an AVP-like substance(s). These include the testis, ovary, uterus, adrenal gland, superior cervical ganglion, and thymus [49]. In addition, various areas of the central nervous system have also been shown to produce AVP [48]. These multiple sources of AVP are consistent with the multiple physiological functions that have recently been shown to be associated with this hormone.

It has been postulated that there are two types of AVP receptors on cells based on antagonist and agonist actions of AVP analogs on renal cells versus hepatic cells [50]. The hepatic cell receptor is designated as V_1, while the renal cell receptor is designated as V_2. Binding of AVP to V_1 receptors (vasopressor receptors) stimulates Ca mobilization and breakdown of phospholipids such as phosphoinositides, while binding to V_2 receptors (antidiuretic receptors) stimulates adenylate cyclase activity and production of cAMP [50].

AVP plays an important role in positive regulation of IFNγ production by providing a helper signal [51, 52]. The helper cell signal for IFNγ production is mediated by the helper T cell lymphokine IL-2 [15]. The experimental approach demonstrating AVP replacement of the IL-2 helper signal for IFNγ production in mouse spleen cell cultures was to first remove helper cells or helper cell function by treatment of cultures with appropriate monoclonal antibodies and complement or with the metabolic inhibitor mitomycin C [51]. Competence for IFNγ production was restored by addition of AVP to cultures along with a T cell mitogen such as staphylococcal enterotoxin A (SEA) [51] (fig. 2). Pressinoic acid, the six amino acid cyclic ring of AVP, appears to be the critical structure for the AVP helper signal, since it restored competence for IFNγ production at a concentration similar to that of AVP. Oxytocin, which has isoleucine in place of phenylalanine in position 3 of the ring, had to be added at tenfold greater concentration for similar help. Isoleucine pressinoic acid, the six amino acid ring structure of oxytocin, was less effective than oxytocin in providing the helper signal. Thus, the six amino acid ring structure of AVP is important for providing the helper signal to lymphocytes for induction of IFNγ with a possible minor role for the three C-terminal amino acids. The classical functions of AVP such as the antidiuretic and vasopressor activities require the three C-terminal amino acids in addition to an intact pressinoic acid ring [53]. This suggests that the AVP receptor on lymphocytes is novel.

In order to better characterize the AVP lymphocyte receptor, V_1 and V_2 receptor antagonists and agonists were examined for their effects on AVP function and binding to lymphocytes. The V_1 antagonist [1-(β-mercapto-β,β-cyclopentamethylene propionic acid), 2-(O-methyl)tyrosine]AVP was a potent competitor of ^{3}H-AVP binding on both liver and spleen cell membranes [54]. In contrast, pressinoic acid did compete with ^{3}H-AVP for binding to the

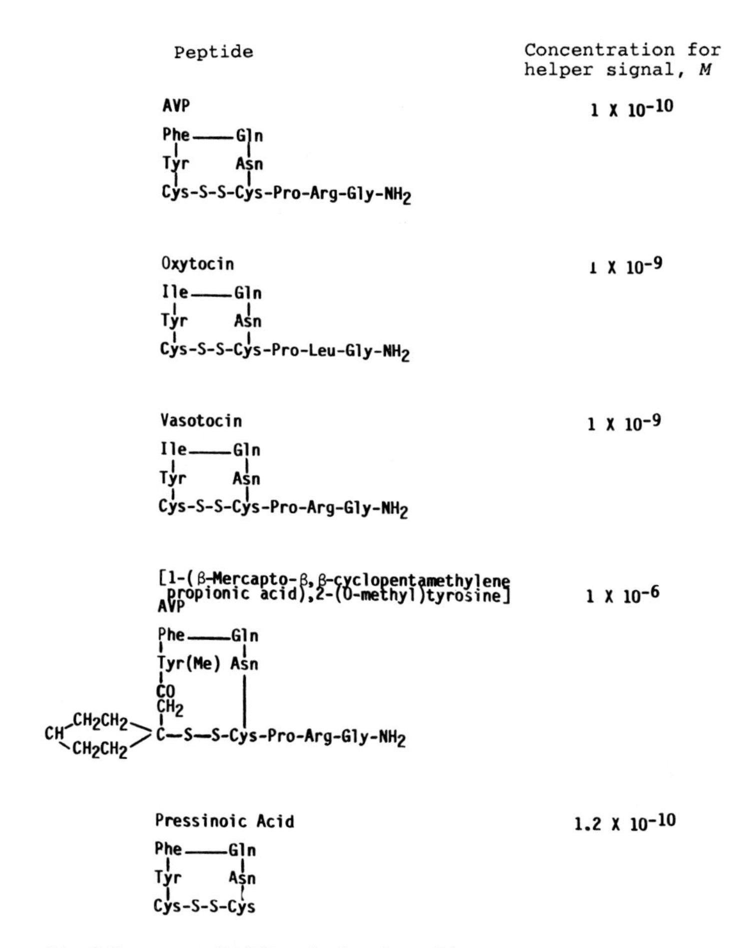

Fig. 2. Structure of AVP and related peptides.

V_1 receptor on spleen cell membranes, but not liver. This suggested that the lymphocyte AVP receptor was unique, yet related to that on liver cells. Functionally, of the several V_1 and V_2 agonists and antagonists examined, only the V_1 antagonist [1-(β-mercapto-β, β-cyclopentamethylene propionic acid), 2-(O-methyl)tyrosine]AVP blocked the AVP helper signal for IFNγ production [52]. Importantly, another V_1 antagonist [1-(β-mercapto-β,β-cyclopentamethylene propionic acid), 2-*D*-(O-ethyl)tyrosine, 4-valine]AVP had no effect on AVP function. IL-2 help was not affected by any AVP antagonists tested. The findings provide direct and functional evidence that lymphocytes possess specific receptors for AVP that are V_1-like but novel.

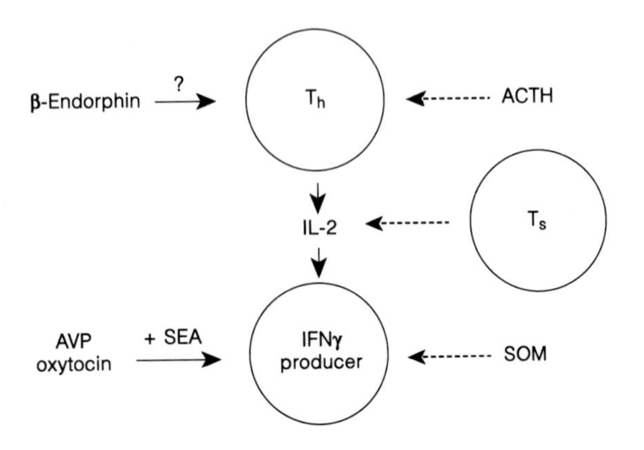

Fig. 3. IFNγ production is regulated by the dynamic interaction between helper (T_h), suppressor (T_s) and IFNγ producer cells. Helper cells provide IL-2. ACTH blocks helper cell function. Suppressor cells absorb IL-2. AVP can replace IL-2 helper signal. β-Endorphin may enhance IL-2 production, and SOM may decrease IFNγ production. Arrows: ——→ = positive signal; – – –→ = negative signal.

Thus, AVP provides a positive signal for induction of IFNγ [51] as compared to the negative signal of ACTH described previously [13]. Whereas ACTH inhibits IFNγ production by blocking a required helper T cell function [13], AVP provides its positive signal by replacing a required helper T cell function, the production of IL-2. It is not known whether ACTH blocks helper cell function by inhibiting production of IL-2. The current scheme for the cellular interactions that regulate IFNγ production in the murine system and the positive role of AVP in induction, along with the negative role of ACTH, is summarized in figure 3. The model suggests a possible interaction between AVP and ACTH regulation of lymphocyte function that was previously not appreciated.

Thyroid-Stimulating Hormone (Thyrotropin, TSH) and Related Hormones

Structurally, TSH is a 28-kDa glycoprotein that is made up of two noncovalently linked subunits designated α and β. The α-subunit resembles that of several other neuropeptides, namely follicle-stimulating hormone (FSH), human chorionic gonadotropin (hCG), and luteinizing hormone (LH)

[55]. The β-subunit is unique for each of these hormones and is responsible for the specificity of their actions. A conformational determinant requiring regions on both subunits constitutes the receptor-binding portion of TSH [56].

Initial observations of hypothyroidism in thymectomized or athymic mice suggested a relationship between the thyroid and the immune system [57]. In fact, human lymphocytes stimulated with the T cell mitogen SEA were shown to produce immunoreactive TSH [58]. Lymphocyte-derived and pituitary TSH were antigenically related and had the same molecular mass and structure. Production of immunoreactive TSH by the T lymphoblast line MOLT-4 could be induced in a dose-dependent fashion by thyrotropin-releasing hormone (TRH) and its induction inhibited by triiodothyronine (T_3) [59].

In addition to TSH being produced by cells of the immune system, some lymphoid cells express TSH receptors. Monocytes, NK cells and mitogen-stimulated B cells, but not phytohemagglutinin (PHA)-stimulated T cells or resting T and B lymphocytes, have been found to bind TSH [60]. It was suggested that aberrant expression of the TSH receptor on lymphoid cells may act as the trigger for anti-TSH receptor autoantibody production observed in the circulation of patients with Graves' disease. Binding of TSH to mitogen-stimulated B cells was associated with an increase in immunoglobulin production. This is consistent with functional studies where both human and bovine TSH enhanced the in vitro antibody response to the T-dependent antigen SRBC in the mouse spleen cell system at concentrations in the range of 0.1 n*M* [61]. TSH had to be present during the first 24–48 h of 96-hour cultures for enhancement of the antibody response, suggesting that TSH affected an early event in the antibody response. TSH also enhanced the antibody response to the *Brucella abortus*-trinitrophenyl (BA-TNP) antigen, which is a less T-dependent antigen than SRBC, yet TSH enhancement of antibody production still required the presence of T cells [62]. Thus, in the immune system, TSH functions to augment both T-dependent and T-independent antibody production. TSH secretion by, and activity on, immune cells may have profound positive regulatory influences on resultant antibody responses.

The structural similarity of TSH with FSH, hCG, and LH would suggest that these hormones should also be investigated both as possible products and as potential regulators of lymphocyte function. Immunoreactive hCG is produced by lymphocytes in the mixed lymphocyte reaction, and it has been proposed that since the blastocyst is implanted in a lymphocyte infiltrate, lymphocyte-derived hCG may play an important role in enhancing implantation of the allogeneic fetus [63]. hCG has been shown to inhibit cytotoxic T cell and NK cell killing, T cell mitogenesis and mixed lymphocyte reactions, potentially via the induction of suppressor cell activity [64, 65]. This would indicate that diversity at the major histocompatibility complex and the subsequent production of hCG would suppress the maternal immune response thereby facilitating fetal survival.

Substance P

Substance P is an eleven amino acid peptide that is a member of the tachykinin family of sensory neuropeptides, which cause tachycardia by lowering peripheral blood pressure. As a group the tachykinins are characterized by C-terminal amino acid sequence Phe-X-Gly-Leu-Met-NH_2, where X is an aromatic or aliphatic amino acid [66]. Substance P is widely distributed in the central nervous system and peripheral sensory nerves that innervate organs and tissues such as the gastrointestinal tract, respiratory tract, visual system, skin, and, relevant to the human system, lymph nodes and spleen [67]. The wide tissue distribution of substance P is consistent with the central role that it is thought to play in pain [67]. Other effects of substance P on nonneural tissues include contraction of intestinal smooth muscle, arteriolar vasodilatation, increased secretion by salivary glands and nasal epithelium, and alteration of microvascular permeability [67]. The latter effect may facilitate cell traffic to local sites of inflammation.

Three tachykinin receptors have been cloned and classified on the basis of relative potencies of various mammalian tachykinins. Human peripheral T cells and the human lymphoblastoid cell line IM-9 express a substance P receptor similar to the NK-1 prototype receptor [68]. Substance P binds to the lymphocyte receptor with $K_d = 4\text{–}8.7 \times 10^{-10}$ *M*, with 7,000 and 25,000 receptors/cell, for T and B cells, respectively [68–70]. Cross-linking of labeled substance P to IM-9 cells and membranes under conditions which minimized proteolysis detected a 58-kDa protein [71]. Interaction of substance P with its receptor stimulated PI turnover in IM-9 cells, and the nonhydrolyzable GTP analog GppNHP inhibited substance P binding [70]. The substance P receptor appears to be coupled to a guanine nucleotide-binding (G) protein and belongs to a family of G protein-linked receptors with seven hydrophobic transmembrane domains. Through the use of analogs and truncated peptides, it has been suggested that the C-terminal portion of substance P is required for binding to its receptor and that the N-terminal basic amino acids are required for cell activation [72].

Substance P has been shown to modulate the activity of cells involved in inflammation. Substance P is a chemoattractant for immune cells and can influence their metabolism and stimulate cell proliferation. On rat tissue mast cells, substance P bound to specific receptors to induce histamine and leukotriene release at micromolar concentrations [73]. The hypersensitivity effects of substance P such as vasodilatation and alteration of microvascular permeability appear to be mediated by substance P-induced factors such as histamine and the leukotrienes. In contrast to the activation of tissue mast cells to release mediators, substance P has no effect on basophils [73]. Thus, the hypersensitivity effects of substance P are cell-specific.

Substance P has also been reported to stimulate chemotaxis and lysosomal enzyme secretion by rabbit neutrophils [74]. This function of substance P may be dependent on the C-terminal substituent peptide and operate via the receptor for the synthetic chemotactic peptide N-formylmethionyl-leucyl-phenylalanine (fMLP). The fact that approximately 1,000 times more substance P than fMLP was required for competitive binding against fML[^{3}H]P and for chemotaxis may suggest that substance P enhanced neutrophil chemotaxis via cross-reactivity with the fMLP receptor rather than through interaction with a primary substance P receptor. Substance P has been shown to stimulate neutrophil phagocytosis as well [74–76]. The substance P N-terminal peptide is structurally similar to the phagocytosis-stimulatory tetrapeptide tufsin [75], and was as effective as intact substance P in activation for phagocytosis. In this case, receptor competition studies suggested that substance P and its N-terminal peptide stimulated phagocytosis by binding to the tufsin receptor.

Substance P has a stimulatory effect on lymphocytes with respect to both T cell proliferation and immunoglobulin production by B cells [77, 78]. Approximately 10% of $CD8^+$ T cells and 18% of $CD4^+$ T cells express specific substance P receptors [79]. The receptor on the helper/inducer subpopulation may be related to their ability to stimulate immunoglobulin synthesis. In the human and mouse systems, substance P stimulated and/or enhanced T cell mitogen-induced T cell proliferation by 60–70% as reflected by incorporation of [^{3}H]thymidine ([^{3}H]TdR) [77]. Unlike neutrophil chemotaxis which does not necessarily involve the NK-1 receptor, the lymphocyte proliferative effects of substance P reside in the C-terminal end of the molecule as the octapeptide substance P_{4-11} was as effective as intact substance P in inducing proliferation. Thus, various functions can apparently involve different parts of the substance P molecule. In addition to T cell proliferative effects, substance P has also been reported to stimulate immunoglobulin production of the IgA and IgM but not IgG class in an in vitro mouse system [78]. Thus, substance P has a positive regulatory effect on both T cell and B cell function.

Monocytes respond chemotactically to substance P and its activity could not be blocked by fMLP antagonists [80]. In the activation of guinea pig macrophages, substance P could stimulate production of the arachidonic acid metabolites, leukotriene C_4, and prostaglandin E, as well as O_2^-, hydrogen peroxide and the cytokines, IL-1, TNFα, and IL-6 [76, 81]. Cytokine production may further affect lymphocyte function, for example, IL-6 production could lead to enhanced immunoglobulin synthesis by B cells. Substance P also augmented macrophage phagocytosis with the release of lysosomal enzymes [76]. Since substance P is widely distributed in nervous tissue, it has been suggested that it may play a role in neurologic disorders such as allergic neuritis, multiple sclerosis, and polyneuritis, which are characterized by extensive macrophage infiltration at focal sites of ongoing tissue damage [76]. Further, substance P may exacerbate rheumatoid arthritis by stimulation of macrophage-like synoviocytes in the joints [82]. Homology has been observed between the amyloid β protein deposited in the brains of patients with Alzheimer's disease and the tachykinins [83]. Substance P could reverse both early neurotrophic and late neurotoxic effects of an amyloid β protein peptide. Therefore, substance P may have neuronal effects opposite to amyloid β protein, but recent reports indicate that amyloid β protein-related peptides do not interact at the substance P receptor itself [84].

Vasoactive Intestinal Peptide

Vasoactive intestinal peptide (VIP) is a 28-amino acid neurotransmitter with partial sequence homology to glucagon, secretin, and growth hormone releasing hormone [66]. It was initially isolated from porcine intestinal tissue and was found to act as a vasodilator of enteric blood vessels. VIP is found in the autonomic and central nervous system, and in both cholinergic and peptidergic neurons innervating several tissues such as the intestine [85], pituitary [86], heart [87], kidney [88], skin [89], thymus [90], and spleen [91]. A wide range of physiological actions have been ascribed to VIP: (a) stimulation of water secretion by the intestine and pancreatic acinar cells [92]; (b) stimulation of prolactin secretion by the pituitary [93]; (c) stimulation of glycogenolysis by liver [94] and brain [95]; (d) inhibition of somatostatin release by the hypothalamus [96]; (e) relaxation of smooth muscles in the trachea [97], intestines [92] and urogenital tract [98], and (f) dilatation of small blood vessels [66].

Human polymorphonuclear leukocytes, eosinophils, and rat mast cells have been demonstrated to contain immunoreactive VIP [99, 100]. In the past it has not been clear whether such cells actually synthesize VIP or simply store VIP obtained from peptidinergic neurons innervating immunologic tissue. Recently, it has been demonstrated that rat basophil leukemia cells produce, store, and release upon appropriate stimulation distinct forms of VIP not found in other tissues [101]. These cells do not appear to have pathways necessary to produce intact VIP_{1-28}. Instead the predominant form produced here is VIP_{10-28} with the C-terminal asparagine as a free acid. Longer forms of N-terminally extended VIP have also been detected; however their amino acid sequences remain to be determined. It will be interesting to determine the effects of these factors on the immune system.

VIP has been reported to have an inhibitory effect on a number of immune responses. It inhibits mitogen-stimulated proliferation of mouse lymphocytes isolated from spleen, Peyer's patches, and mesenteric lymph nodes [102]. T lymphocytes seem to be the primary target of VIP-mediated inhibition since proliferative responses to T cell mitogens such as Con A and PHA were affected, but proliferation induced by the B cell mitogen LPS was not inhibited. The findings are consistent with the observation that T cells, but not B cells, have high affinity receptors for VIP [103].

In addition to effecting proliferation in response to mitogenic stimulation, VIP has also been demonstrated to modulate cytokine production. For example, VIP has been shown to inhibit the generation of IL-2 and IFNγ from mitogen-stimulated murine and human lymphocytes [104].

Immunoglobulin production, in particular IgA, was also shown to be affected by VIP [78]. IgA production by Con A-stimulated lymphocytes from spleen and mesenteric lymph nodes was slightly increased by VIP (30 and 20%, respectively) whereas IgA production by Peyer's patches-derived lymphocytes was significantly increased (70%). Interestingly, IgM production in Peyer's patches was increased (80%) by VIP but was unaffected in spleen and lymph nodes. IgA is the predominant immunoglobulin produced in Peyer's patches, and IgA-producing lymphocytes are more abundant in these tissues than in lymph nodes or spleen [105, 106]. The site-specific VIP effects on immunoglobulin production can perhaps be explained by differences in T cell subpopulations in these organs.

VIP has been reported to affect human NK cell activity [107], inhibiting tumoricidal activity at VIP concentrations ranging from 10^{-7} to 10^{-10} *M*. Although these concentrations are higher than those normally found in peripheral blood (10^{-11} *M*), concentrations in the vicinity of VIP-secreting neurons have been estimated to be 10^{-9}–10^{-10} *M* [108]. VIP effects on macrophage-mediated tumoricidal activity were also investigated, and although VIP alone did not affect tumor cell killing, it potentiated the suppressive effects of noradrenaline [109]. More recently others have demonstrated that VIP and its related peptides derived from preproVIP, peptide histidine methionine (PHM-27) and peptide histidine valine (PHV-42), inhibited NK activity. The order of potency being VIP > PHV-42 > PHM-27. Interestingly, the VIP antagonist [*p*-chloro-Phe^6,Leu^{17}]-VIP at 10^{-8} *M* inhibited the effects of VIP but not PHM-27 or PHV-42, thus suggesting the existence of different receptors for prepro-VIP-derived peptides [110].

Evidence also suggests that VIP can modulate lymphocyte traffic. Egress of lymphocytes from popliteal lymph nodes is inhibited by VIP treatment [111]. In another report, mouse T lymphocytes were pretreated with VIP for 18 h, reinjected into syngeneic subjects and their subsequent in vivo localization was monitored [112]. Lymphocytes pretreated with VIP showed a reduced ability to migrate back to mesenteric lymph nodes and Peyer's patches. Lymphocyte distribution in spleen and blood was unaffected. Lymphocytes pretreated with VIP and washed of excess VIP showed decreased specific binding of ^{125}I-VIP with Scatchard analysis indicating a marked reduction in the number of binding sites/cell. The altered expression of VIP receptors was not associated with a concomitant alteration in expression of Thy-1, Lyt-1, or Lyt-2 phenotypic markers. Binding of VIP to its lymphocyte receptor appeared to result in a decreased expression of the VIP receptor. Down-regulation of the VIP receptor may, in turn, affect the interaction of lymphocytes with lymphoid tissue, causing decreased infiltration into mesenteric lymph nodes and Peyer's patches.

Receptors for VIP have been shown to exist on a number of immune effector cells, including human peripheral blood lymphocytes [113], mouse T lymphocytes [102], human mononuclear cells [103] and the human lymphoblastic cell line Molt-4B [114]. Human lymphocytes possess both high and low affinity VIP receptors, with the high affinity receptor having a $K_d = 4.7 \times 10^{-10}$ *M* and the low affinity receptor having a $K_d = 8 \times 10^{-8}$ *M*. These lymphocyte receptors are similar in specificity and K_d to the high and low affinity VIP receptors found in brain, endocrine, and intestinal tissue [115].

There appears to exist a differential expression of receptors on the various lymphocyte phenotypes [116]. For example, human T cell-enriched PBL preparations demonstrated enhanced binding of ^{125}I-VIP as compared to unseparated or T cell-depleted preparations. Enriched populations of CD4 and CD8 cells showed even higher binding capacity, however CD4 was greater than CD8 in this regard. Also, ^{125}I-VIP specifically bound to large granular and B lymphocytes, with the latter having the lowest binding capacity.

Cross-linking studies have been performed to characterize the VIP receptor. Through the use of various cross-linkers, a 47-kDa protein has been isolated from Molt-4B lymphoblasts and GH_3 pituitary cells. Rat frontal cortex and pancreatic acinar cells have also been demonstrated to have a 47- to 48-kDa cross-linked protein. Liver and intestinal epithelium, on the other hand, demonstrated two cross-linked species: a high molecular weight (73–86 kDa), high affinity protein and a low molecular weight (30–36 kDa), low affinity protein.

Further studies were performed on Molt-4B lymphoblasts and GH_3 pituitary cells to determine the functionality of the VIP receptor. VIP caused a cascade of intracellular events in these cells: activation of adenylate cyclase, accumulation of cAMP, activation of cAMP protein kinase, and phosphorylation of specific cellular proteins [92, 117].

It has been proposed that G proteins modulate VIP receptor interaction with adenylate cyclase [115]. G proteins act as transmembrane signals and interact with the catalytic subunit of adenylate cyclase to either inhibit (via G_i) or stimulate (via G_s) enzymatic activity [118]. Interaction of VIP with its high-affinity receptor results in G protein, G_s, stimulation of adenylate cyclase. Somatostatin, a known inhibitor of cAMP formation in adenohypophyseal cells, antagonizes VIP effects, perhaps by facilitating the interaction of G protein, G_i, with either G_s or adenylate cyclase [115].

An interesting model emerges from the reported data on VIP and its immunomodulatory effects. VIP synthesized and secreted by either immune effector cells or neurons innervating lymphoid tissue can act on VIP receptor-bearing cells (such as T lymphocytes), resulting in inhibition of proliferative responses, IgA production, and migration of lymphocytes to lymphoid tissues such as mesenteric lymph nodes and Peyer's patches. These inhibitory effects may be the result of activation of adenylate cyclase. The stimulation of adenylate cyclase may be mediated by G proteins, in particular G_s, which binds to the catalytic subunit of adenylate cyclase as a result of the interaction of VIP with its membrane receptor. The subsequent accumulation of cAMP causes the activation of a cAMP-dependent protein kinase which phosphorylates specific cellular proteins. The function of proteins phosphorylated as a result of VIP-receptor interaction has not yet been determined, although they may possibly be involved in VIP immunomodulatory effects.

Growth Hormone

Growth hormone (GH), also known as somatotropin, is a 191-amino acid hormone of the adenohypophysis [119] which has structural similarities to prolactin (PRL). Probably its best characterized activity is the promotion of growth of bone, cartilage, and soft tissues. Although GH secretion is maximal at puberty and declines after about 30 [120], detectable levels are found throughout the remainder of adulthood. This suggests that this hormone has other functions in addition to promotion of growth. In this respect, GH may be important for the maintenance of lean body mass. GH effects are mediated in part by insulin-like growth factor I (IGF-I) or somatomedin, which is induced by GH stimulation of hepatocytes.

Recently, investigators have determined that the age-related changes in body composition of elderly men (61- to 81-year olds) with low plasma IGF-I concentration could be altered by the administration of recombinant biosynthetic human GH [121]. Treatment consisted of administering about 0.03 mg of GH/kg body weight subcutaneously 3 times a week. Plasma IGF-I was used as an indirect measurement of GH secretion. The results of this study were that IGF-I levels increased to a range similar to that of younger men. Physiologically, a significant increase in lean body mass (8.8%), average density of lumbar vertebra (1.6%), and sum of skin thickness (7.1%) was observed. In contrast, the adipose tissue mass decreased by 14.4%. It was concluded that diminished secretion of GH as indirectly measured by plasma IGF-I levels is responsible, at least in part, for the changes associated with old age (lean body mass, adipose-tissue mass, and thinning of skin).

Given the likelihood that the decline of production of GH with age may play a role in the aging process, it is important to determine possible extrapituitary sources of GH. It is of particular interest that lymphocytes have been identified in the ectopic secretion of GH [122–124]. Nonstimulated rat lymphocytes secrete immunoreactive GH (irGH), which is similar in structure to pituitary-derived GH [122–124]. Lymphocyte-derived irGH is secreted spontaneously in culture while pituitary GH is secreted under the control of GHRH.

Evidence has accumulated indicating that GH not only regulates growth of body tissues but can control important immune functions. One of the first lines of evidence to demonstrate how adenohypophyseal hormones are related to immunoregulation was that hypophysectomy caused thymic atrophy [125]. The functional significance of this finding has probably not been appreciated until recently. Studies have now shown that hypophysectomized rats were unable to exhibit an immune response [126]. Treatment with GH or PRL actually restored immunocompetence, whereas other pituitary hormones had no effect [127]. These observations are consistent with reports that lymphocytes express specific receptors for GH [128–130] and PRL [131, 132].

Thymic atrophy is seen in experimental hypophysectomy or advanced aging. Previously, investigators have tried to determine the deleterious consequences of aging on the immune function by implanting GH and PRL secretory GH_3 pituitary adenomas cells into aging rats (18–24 months old). Implantation here resulted in partial regeneration of thymic tissue and partial restoration of T cell competence for proliferation [133]. The production of IL-2 was significantly increased in these rats. Since previous observations indicated that GH_3 implantation caused a dramatic rise in plasma GH and plasma PRL [134, 135], the authors surmised that these hormones may have been responsible for the restoration of T cell responses seen in these animals. The findings do not exclude the possibility of other GH_3-derived hormones being important here. Further, treatment of these engrafted animals with GH- and/or PRL-specific antibodies could possibly provide information to clarify the role that GH and PRL play in the modulation of immunoregulation [136]. However, in support of these findings, exogenous GH injection of aging rats resulted in augmented spleen cell proliferation and NK cell activity [136]. Yet no morphologic alterations were seen in the thymus of the animals. Studies determining the combined effects of coadministration of exogenous GH and PRL on restoration of immune function would be of interest.

GH is also responsible for modulating other immunologic activities such as augmenting cytolytic activity of T cells, antibody synthesis, GM-CSF-dependent granulocyte differentiation, TNFα production, superoxide anion generation from peritoneal macrophages of hypophysectomized mice and NK activity [137].

The regulatory mechanisms governing irGH secretion and its effects on lymphocytes have recently been examined [138]. Administration of an antisense oligonucleotide of GH mRNA to lymphocytes inhibited T and B cell proliferation. Antisense, but not sense, oligonucleotides were able to significantly inhibit irGH production and [^{3}H]TdR incorporation in both resting and Con A-stimulated lymphocytes. These findings suggest that irGH is important for lymphocyte proliferation and that this activity may be facilitated through an 'autocrine/paracrine mechanism'. Importantly, lymphocyte-derived GH appears to be necessary for interleukin-induced lymphocyte growth, although it is not directly mitogenic.

Prolactin

PRL is a 170-amino acid long neuropeptide secreted from the adenohypophysis [119]. It has been demonstrated to participate in a multitude of physiologic activities [139], however it is best known for its role in the initiation and maintenance of lactation [119]. Also, it is ubiquitous in nature in that it has been identified in all vertebrates [139]. Several lines of evidence have accumulated to suggest that PRL has important immunoregulatory activities. Factors that cause hypoprolactinemia are generally associated with inducing immunoincompetence [140, 141].

One of the first lines of evidence to suggest the importance of PRL as well as GH on immune function was the observation that hypophysectomy resulted in the regression of the thymus [125] and, more recently, has been shown to result in loss of immune competence [126]. Treatment with PRL restored immune competence [127]. In another study, chemically induced hypoprolactinemia in mice using bromocriptine, a dopamine type 2 agonist, resulted in decreased macrophage-mediated tumoricidal activity, inhibition of mitogen-driven proliferation of splenic lymphocytes, and increased mortality from intraperitoneal challenge of the bacterial pathogen *Listeria monocytogenes* [140]. Coadministration of ovine PRL reversed these effects. Similarly, CQP 201-403, a more potent, longer acting dopamine type 2 agonist, was used to suppress PRL levels in a rat cardiac allograft model [141]. Maximal dose administration caused a significant increase in graft survival as compared with controls. Concurrent administration of a low dose of ciclosporin resulted in a doubling of graft survival time compared to ciclosporin alone. Increased circulating levels of PRL have been seen in patients experiencing acute cardiac allograft rejection [142]. This may be due to the fact that ciclosporin, which was used to suppress graft rejection, can act as an antagonist to PRL binding to the PRL receptor on lymphocytes [131]. Hence, one mechanism of ciclosporin-induced immunosuppression may be to block the immunostimulatory effects of PRL on lymphocytes. As alluded to earlier, therapy that is directed at reducing circulating levels of PRL may facilitate the use of subtoxic and yet immunosuppressive doses of immunosuppressive drugs such as ciclosporin to prevent allograft rejection [141].

There is evidence that PRL may act as a lymphocyte growth factor [143–146]. Increased expression of growth-related genes, such as ornithine decarboxylase and c-myc, have been shown to occur in cells treated with PRL [143, 145]. The mechanism of expression of the above genes appears to involve activation of protein kinase C by PRL, particularly activation at the level of the nuclear membrane [144].

It has been established that human PRL is normally secreted from the adenohypophysis, chorion or endometrium. Ectopic secretion of PRL is rarely seen with certain carcinomas [147]. Intracellular PRL has been detected in certain nonpituitary malignant cell lines, however secretion of PRL was not documented [148]. More recently, PRL has been shown to be secreted from a human B lymphoblastoid cell line variant designated as IM-9-P3 [149]. PRL synthesized by these cells was indistinguishable from human pituitary PRL, however the mRNA species expressed differed from that of pituitary origin in that it was approximately 150 bases longer than expected. The authors speculated that there was an elongation of the 5′ and/or 3′ untranslated regions of IM-9-P3 derived PRL mRNA. Further confirmation of the identity of this PRL is demonstrated by the fact that the immunoaffinity-purified lymphocyte PRL is of the same molecular weight as human pituitary PRL [149].

A 46-kDa PRL-like molecule has been shown to be secreted by mitogen-driven rodent lymphocytes as compared to a 25-kDa pituitary PRL [150]. One group has reported that a PRL-like mRNA from proliferating splenocytes is larger than other species of PRL-like mRNAs [124]. The size of mRNAs from PRL-like molecules from nonlymphoid tissues are generally in the size range of 1 kb [151–153]. Thus, in addition to PRL, there are also PRL-like molecules from lymphocytes as well as nonlymphoid tissues.

The findings presented above would suggest that PRL or PRL-like molecules play an important role in positive regulation of lymphocyte functions. Future studies should address the relationship of PRL activation of lymphocytes to that of interleukins such as IL-2.

Somatostatin

Somatostatin (SOM) is a 14-amino acid peptide found in the hypothalamus and central and peripheral nervous system. It is also located in the gastrointestinal tract and pancreatic islets. Its activities are pervasive in that it is responsible for inhibiting the secretion of GH, VIP, TSH, glucagon, insulin, secretin, gastrin, and cholecystokinin. Whereas GH and PRL have immunoenhancing capabilities, SOM has potent inhibitory effects on immune responses.

SOM has been shown to significantly inhibit Molt-4 lymphoblast proliferation and PHA stimulation of human T lymphocytes at 10^{-12} *M* [154]. Also, nanomolar concentrations were able to inhibit the proliferation of both spleen- and Peyer's patches-derived lymphocytes [78]. Interestingly, intravenous infusion of SOM in patients to treat active duodenal ulcers caused a significant reduction in proliferative responses of mitogen-stimulated circulating lymphocytes [155]. Other immune responses, such as SEA-stimulated IFNγ secretion [156] (fig. 3), endotoxin-induced leukocytosis [157] and colony-stimulating activity release [158] are also inhibited by SOM. These functional data suggest that these responsive tissues express receptors for SOM.

Several studies have demonstrated that cells of the immune system express receptors for SOM [159, 160]. Recently, one group has described the existence of distinct subsets of SOM receptors on the Jurkat line of human leukemic T cells and U266 IgE-producing human myeloma cells [159]. They showed that these cells have both high and low affinity receptors with K_d values in the p*M* and n*M* range respectively. The authors speculate that two subsets of receptors may account for the biphasic concentration-dependent nature of the effects of SOM in some systems [154, 161].

Thus, pituitary hormones such as GH and PRL have immunoenhancing effects on T cells, causing increased responses to mitogen stimulation and lymphokine production, while SOM, a potent hypothalamus-derived inhibitor of GH release, inhibits T cell proliferation. An interesting loop involving positive and negative signals emerges concerning GH, PRL and SOM regulation of immune function.

Second Messenger Signals for Neuroendocrine Hormones

The ability of neuroendocrine hormones to replace lymphokine function, such as AVP replacing the IL-2 requirement in IFNγ production,

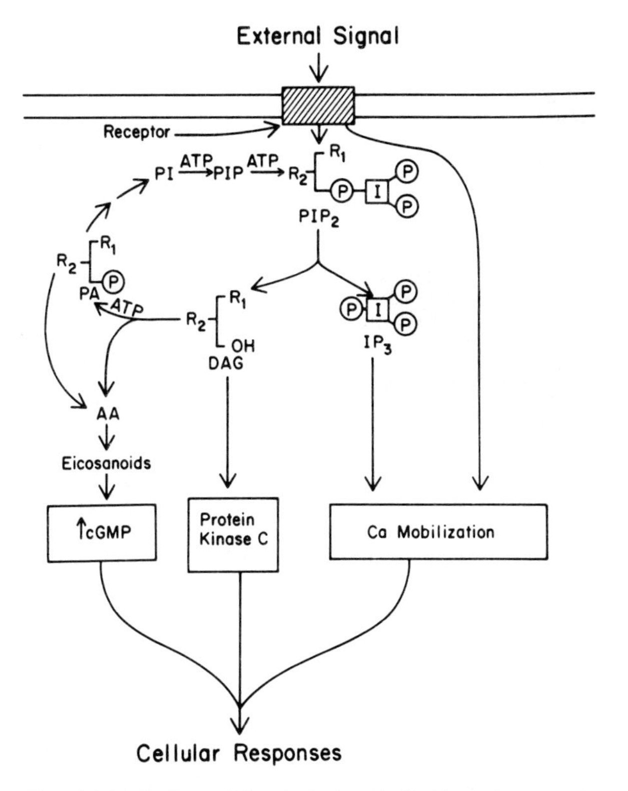

Fig. 4. Metabolism of inositol phospholipids and generation of second messenger signals for various cellular responses. Phosphatidylinositol-4,5-bisphosphate (PIP_2) is thought to be the immediate target for signal-initiated hydrolysis. PI = phosphatidylinositol; PIP = phosphatidylinositol-4-phosphate; IP_3 = inositol-1,4,5-trisphosphate; DAG = 1,2-diacylglycerol; PA = phosphatidic acid; cGMP = cyclic GMP; I = inositol; R_1 and R_2 = fatty acid acyl groups; P = phosphoryl group.

suggests the possibility of common second messenger signals. AVP has been shown to stimulate phospholipid turnover in several systems [162–164]. Membrane phospholipids, particularly phosphatidylinositol phosphates, are believed to play a central role in generation of second messenger signals as a result of ligand-membrane receptor interaction [165]. Interaction of ligand with the membrane receptor results in the hydrolysis of phosphatidylinositol-4,5-bisphosphate (PIP_2) by phospholipase C (fig. 4). PIP_2 exists in equilibrium with phosphatidylinositol (PI) and phosphatidylinositol-4-phosphate (PIP). It is cleaved to produce inositol-1,4,5-trisphosphate (IP_3) and

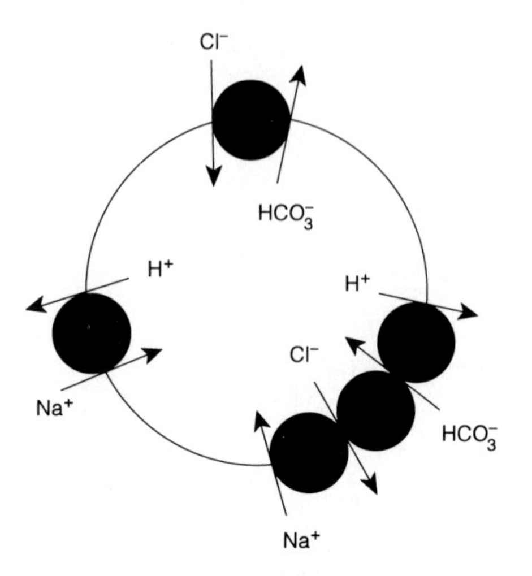

Fig. 5. Ion exchanges involved in regulation of intracellular pH, pH_i, in renal mesangial cells stimulated with AVP in the presence of bicarbonate. AVP stimulates sodium-independent Cl^-/HCO_3^- exchanger (top) more than it stimulates the sum of the Na^+/H^+ and sodium-dependent Cl^-/HCO_3^- exchangers (two toward the bottom) thus lowering pH_i.

diacylglycerol (DAG). With the production of arachidonic acid from DAG or from phosphatidic acid, we have a trifurcation of second messenger signal generation for regulation of cellular responses via PIP_2 metabolism.

The IP_3 messenger mediates its effects via stimulation of a large and early release of calcium from the endoplasmic reticulum [165]. The increase in the intracellular free calcium concentration is thought to stimulate protein phosphorylation, which ultimately leads to the appropriate cellular response.

DAG, representing another second messenger pathway, acts by stimulating protein kinase C phosphorylation [166]. Protein kinase C in the phosphorylated state may play a role in stimulation of cellular activity via activation of three distinct intracellular pH (pH_i) regulatory mechanisms that are active in the presence of bicarbonate (fig. 5) [167]. One involves Na^+/H^+ exchange and may be the principal mechanism for extruding acid from cells. The second is a sodium-dependent Cl^-/HCO_3^- exchange. The third mechanism is a simple Cl^-/HCO_3^- exchanger, and is independent of sodium. AVP causes acidification in the presence of bicarbonate as a result of stimulation of the sodium-independent Cl^-/HCO_3^- exchanger (which lowers pH_i) more than it stimulates the sum of the Na^+/H^+ and the sodium-dependent $Cl^-/$

HCO_3^- exchangers (both of which raise pH_i). AVP stimulation of these three acid-base transporters enhances the ability of the cell to regulate pH_i.

Arachidonic acid, the third second messenger of PIP_2 metabolism, is produced from DAG, and other phospholipids [168]. There is evidence that arachidonic acid can serve directly as a second messenger for some cellular functions and that its lipoxygenase and cyclo-oxygenase products can also serve as second messenger signals. Arachidonic acid and structurally related unsaturated fatty acids have recently been reported, for example, to directly activate protein kinase C, suggesting that DAG is not the only activator of this enzyme [169]. Arachidonic acid and its lipoxygenase products (leukotrienes, hydroxyeicosatetraenoic acid (HETES, etc.) have been reported to activate guanylate cyclase activity and thus increase the cyclic GMP level of cells [170]. Thus, some of the second messenger signals associated with this fatty acid may result from induction of cyclic GMP.

Each of the second messenger signal pathways described above is a potential candidate for mediation of the AVP helper signal for IFNγ production. Possible involvement of the arachidonic acid signal pathway is suggested by the ability of arachidonic acid, its lipoxygenase products (leukotrienes and HETES), and dibutyryl cyclic GMP to completely mediate the helper signal for IFNγ production [171–173]. The synthetic DAGs 1-oleoyl-2-acetylglycerol and sn-1,2-dioctanoylglycerol, which activate protein kinase C in intact cells, could not provide the helper signal [172]. Thus, induction of arachidonic acid with possible activation of guanylate cyclase may provide important second messenger signals for mediation of the helper signal for IFNγ production, while activation of protein kinase C does not appear to be sufficient for such help. Both the release of arachidonic acid and its inductive helper signal for IFNγ production are calcium-dependent, since competitive ionic antagonists of calcium block IFNγ production [174]. In this regard further studies need to be carried out to determine the role of IP_3 and mobilization of intracellular calcium in IFNγ induction. Finally, the second messenger signals produced in phospholipid turnover should be examined for other positive neuroendocrine hormone signals associated with immune function.

There is evidence that G proteins are involved in the antagonistic effects that VIP and SOM have on lymphocyte function [115]. As indicated earlier, VIP interaction with its receptor initiates G protein, G_s, stimulation of adenylate cyclase. SOM antagonizes VIP action by facilitating interaction of an inhibitory G protein, G_i, with either G_s or adenylate cyclase [115]. The G proteins have also been shown to play a role in phospholipid turnover [175], and thus they may play a central role in the generation of a variety of second messenger signals.

Conclusions

There is a functional and structural relationship between the immune and neuroendocrine systems. As presented in this chapter, neuroendocrine hormones such as ACTH, endorphins, enkephalins, AVP, TSH, substance P, VIP, GH, PRL, and SOM can regulate a number of important immune functions. Some of these are: (a) enhancement and suppression of antibody production; (b) enhancement and suppression of lymphocyte cytotoxicity and proliferation; (c) modulation of lymphokine production and mediation of lymphokine function; (d) initiation of hypersensitivity events and possibly some immune complex diseases; (e) modulation of macrophage and neutrophil functions, and (f) enhancement of primary lymphoid organ function such as thymic regeneration. The fact that lymphocytes can produce many of the neuroendocrine hormones that act on the immune system also makes these substances lymphokines by definition, and provides evidence that their regulation of immune functions may be part of the natural immunoregulatory repertoire.

References

1 Blalock JE: Peptide hormones shared by the neuroendocrine and immunologic systems. J Immunol 1985;135:858s–861s.
2 Smith EM, Galin S, LeBoeuf RD, Coppenhaver DH, Harbour DV, Blalock JE: Nucleotide and amino acid sequence of lymphocyte-derived corticotropin: Endotoxin induction of a truncated peptide. Proc Natl Acad Sci USA 1990;87:1057–1060.
3 Douglas J, Civelli O, Herbert E: Polyprotein gene expression: Generation of diversity of neuroendocrine peptides. Annu Rev Biochem 1984;53:665–715.
4 Blalock JE, Smith EM: Human leukocyte interferon: Structural and biological relatedness to adrenocorticotropic hormone and endorphins. Proc Natl Acad Sci USA 1980;77:5972–5974.
5 Smith EM, Blalock JE: Human lymphocyte production of corticotropin- and endorphin-like substances: Association with leukocyte interferon. Proc Natl Acad Sci USA 1981;78:7530–7534.
6 Johnson HM, Smith EM, Torres BA, Blalock JE: Regulation of the in vitro antibody response by neuroendocrine hormones. Proc Natl Acad Sci USA 1982;79:4171–4174.
7 Mishell RI, Dutton RW: Immunization of dissociated spleen cell cultures from normal mice. J Exp Med 1976;126:423–442.

8 Mosier DE, Scher I, Paul WE: In vitro responses of CBA/N mice: Spleen cells of mice with an X-linked defect that precludes immune responses to several thymus-independent antigens can respond to TNP-lipopolysaccharide. J Immunol 1976;117:1363–1369.
9 Clarke BL, Bost KL: Differential expression of functional adrenocorticotropic hormone receptors by subpopulations of lymphocytes. J Immunol 1989;143:464–469.
10 Alvarez-Mon M, Kehrl JH, Fauci AS: A potential role for adrenocorticotropin in regulating human B lymphocyte functions. J Immunol 1985;135:3823–3826.
11 Munn NA, Lum LG: Immunoregulatory effects of α-endorphin, β-endorphin, methionine-enkephalin, and adrenocorticotropic hormone on anti-tetanus toxoid antibody synthesis by human lymphocytes. Clin Immunol Immunopathol 1989;52:376–385.
12 Carr DJJ, Radulescu RT, DeCosta BR, Rice KC, Blalock JE: Differential effect of opioids on immunoglobulin production by lymphocytes isolated from Peyer's patches and spleen. Life Sci 1990;47:1059–1069.
13 Johnson HM, Torres BA, Smith EM, Dion LD, Blalock JE: Regulation of lymphokine (γ-interferon) production by corticotropin. J Immunol 1984;132:246–250.
14 Epstein LB, Cline MJ, Merigan TC: The interaction of human macrophages and lymphocytes in the phytohemagglutinin-stimulated production of interferon. J Clin Invest 1971;50:744–753.
15 Torres BA, Farrar WL, Johnson HM: Interleukin-2 regulates immune interferon (IFNγ) production by normal and suppressor cell cultures. J Immunol 1982;128:2217–2219.
16 Torres BA, Yamamoto JK, Johnson HM: Cellular regulation of gamma interferon production: Lyt phenotype of the suppressor cell. Infect Immun 1982;35:770–776.
17 Wybran J, Appelboom T, Famaey J-P, Govaerts A: Suggestive evidence for receptors for morphine and methionine-enkephalin on normal human blood T lymphocytes. J Immunol 1979;123:1068–1070.
18 Hazum E, Chang K-J, Cuatrecasas P: Specific nonopiate receptors for β-endorphin. Science 1979;205:1033–1035.
19 Gilmore W, Weiner LP: β-Endorphin enhances interleukin (IL-2) production in murine lymphocytes. J Neuroimmunol 1988;18:125–138.
20 Heijnen C, Bevers C, Kavelaars A, Ballieux RE: Effect of α-endorphin on the antigen-induced primary antibody response of human blood B cell in vitro. J Immunol 1986;136:213–216.
21 Hazum E, Chang K-J, Cuatrecasas P: Interaction of iodinated human [*D*-Ala$_2$]β-endorphin with opiate receptors. J Biol Chem 1979;254:1765–1767.
22 Udenfriend S, Kilpatrick DL: Biochemistry of the enkephalins and enkephalin-containing peptides. Archs Biochem Biophys 1983;221:309–323.
23 Morgan EL, McClurg MR, Janda JA: Suppression of human B lymphocyte activation by β-endorphin. J Neuroimmunol 1990;28:209–217.
24 Williamson SA, Knight RA, Lightman SL, Hobbs JR: Effects of beta-endorphin on specific immune responses in man. Immunology 1988;65:47–51.
25 Rowland RR, Tokuda S: Dual immunomodulation by met-enkephalin. Brain Behav Immun 1989;3:171–174.
26 Shavit Y, Lewis JW, Terman GW, Gale RP, Liebeskind JC: Opioid peptides mediate the suppressive effect of stress on natural killer cell cytotoxicity. Science 1984;223:188–190.

27 Flores CM, Hernandez MC, Hargreaves KM, Bayer BM: Restraint stress-induced elevations in plasma corticosterone and β-endorphin are not accompanied by alterations in immune function. J Neuroimmunol 1990;28:219–225.
28 Mathews PM, Froelich CJ, Sibbitt WL, Bankhurst AD: Enhancement of natural cytotoxicity by β-endorphin. J Immunol 1983;13:1658–1662.
29 Kay N, Morley JE, Allen JI: Interaction between endogenous opioids and IL-2 on PHA-stimulated human lymphocytes. Immunology 1990;70:485–491.
30 Faith RE, Liang HJ, Murgo AJ, Plotnikiff NP: Neuroimmunomodulation with enkephalins: Enhancement of human natural killer (NK) cell activity in vitro. Clin Immunol Immunopathol 1984;31:412–418.
31 Foster JS, Moore RN: Dynorphin and related opioid peptides enhance tumoricidal activity mediated by murine peritoneal macrophages. J Leukocyte Biol 1987;42: 171–174.
32 Oleson DR, Johnson DR: Regulation of human natural cytotoxicity by enkephalins and selective opiate agonists. Brain Behav Immun 1988;2:171–186.
33 Carr DJJ, DeCosta BR, Jacobson AE, Rice KC, Blalock JE: Corticotropin-releasing hormone augments natural killer cell activity through a naloxine-sensitive pathway. J Neuroimmunol 1990;28:53–61.
34 Gilman SC, Schwartz JM, Milner RJ, Bloom FE, Feldman JD: β-Endorphin enhances lymphocyte proliferative responses. Proc Natl Acad Sci USA 1982;79:4226–4230.
35 McCain HW, Lamster IB, Bozzone JM, Grbic JT: β-Endorphin modulates human immunity via non-opiate receptor mechanisms. Life Sci 1982;31:1619–1624.
36 Glimore W, Moloney M, Berinstein T: The enhancement of polyclonal T cell proliferation by beta-endorphin. Brain Res Bull 1990;24:687–692.
37 Hemmick L, Bidlack JM: β-Endorphin stimulates rat T lymphocyte proliferation. J Neuroimmunol 1990;29:239–248.
38 Miller GC, Murgo AJ, Plotnikoff NP: Enkephalins: Enhancement of active T-cell rosettes from lymphoma patients. Clin Immunol Immunopathol 1983;26:446–451.
39 Van Epps DE, Saland L: β-Endorphin and [Met]enkephalin stimulate human peripheral blood mononuclear cell chemotaxis. J Immunol 1984;132:3046–3053.
40 Ruff M, Schiffmann E, Terranova V, Pert CB: Neuropeptides are chemoattractants for human tumor cells and monocytes: A possible mechanism for metastasis. Clin Immunol Immunopathol 1985;37:387–396.
41 Brown SL, Van Epps DE: Suppression of T lymphocyte chemotactic factor production by the opioid peptides β-endorphin and [Met]enkephalin. J Immunol 1985;134: 3384–3390.
42 Peterson PK, Sharp B, Gekker G, Brummitt C, Keane WF: Opioid-mediated suppression of interferon-γ production by cultured peripheral blood mononuclear cells. J Clin Invest 1987;80:824–831.
43 Morgano A, Setti M, Poerri I, Barabine A, Lotti G, Indiveri F: Expression of HLA-class II antigens and proliferative capacity in autologous mixed lymphocyte reactions of human T lymphocytes exposed in vitro to α-endorphin. Brain Behav Immun 1989;3:214–222.
44 Brownstein MJ, Russell JT, Gainer H: Synthesis, transport, and release of posterior pituitary hormones. Science 1980;207:373–378.
45 Sawyer WH: Neurohypophyseal hormones. Pharmacol Rev 1961;13:225–277.
46 Rossi NF, Schrier RW: Role of arginine vasopressin in regulation of systemic arterial pressure. Annu Rev Med 1986;37:13–20.

47 Gibbs DM: Vasopressin and oxytocin: hypothalamic modulations of the stress response: A review. Psychoneuroendocrinology 1986;11:131–140.
48 Doris PA: Vasopressin and central integrative processes. Neuroendocrinology 1984; 38:75–85.
49 Clements JA, Funder JW: Arginine vasopressin (AVP) and AVP-like immune reactivity in peripheral tissues. Endocr Rev 1986;7:449–460.
50 Kirk CJ, Creba JA, Hawkins PT, Mitchell RH: Is vasopressin-stimulated inositol lipid breakdown intrinsic to the mechanism of calcium mobilization at V_1 vasopressin receptors? Prog Brain Res 1983;60:405–411.
51 Johnson HM, Farrar WL, Torres BA: Vasopressin replacement of interleukin-2 requirement in gamma interferon production: Lymphokine activity of a neuroendocrine hormone. J Immunol 1982;129:963–986.
52 Johnson HM, Torres BA: Regulation of lymphokine production by arginine vasopressin and oxytocin: Modulation of lymphocyte function by neurohypophyseal hormones. J Immunol 1985;135:773s–775s.
53 Manning M, Olma A, Klis W, Kolodziejczyk A, Nawrocka E, Misicka A, Seto J, Sawyer WH: Carboxy terminus of vasopressin required for activity but not binding. Nature (Lond) 1984;308:652–653.
54 Torres BA, Johnson HM: Arginine vasopressin (AVP) replacement of helper cell requirement in IFNγ production. Evidence for a novel AVP receptor on mouse lymphocytes. J Immunol 1988;140:2179–2183.
55 Teppermann J, Teppermann HM: Metabolic and Endocrine Physiology, ed 5. Chicago, Year Book, 1987, pp 131–132.
56 Costagliola S, Madec A-M, Benkirane MM, Orgiazzi J, Carayon P: Monoclonal antibody approach to the relationship between immunological structure and biological activity of thyrotropin. Mol Endocrinol 1988;2:613–618.
57 Pierpaoli W: Psychoneuroimmunology. New York, Academic Press, 1981, pp 575–606.
58 Smith EM, Phan M, Kruger TE, Coppenhaver D, Blalock JE: Human leukocyte production of immunoreactive thyrotropin. Proc Natl Acad Sci USA 1983;80:6010–6013.
59 Harbour DV, Kruger TE, Coppenhaver D, Smith EM, Meyer WJ: Differential expression and regulation of thyrotropin (TSH) in T cell lines. Mol Cell Endocrinol 1989;64:229–241.
60 Coutelier J-P, Kehrl JH, Bellur SS, Kohn LD, Notkins AL, Prabhakar BS: Binding and functional effects of thyroid stimulating hormone on human immune cells. J Clin Immunol 1990;10:204–210.
61 Blalock JE, Johnson HM, Smith EM, Torres BA: Enhancement of the in vitro antibody response by thyrotropin. Biochem Biophys Res Commun 1984;125:30–34.
62 Kruger TE, Blalock JE: Cellular requirements for thyrotropin enhancement of in vitro antibody production. J Immunol 1986;137:197–200.
63 Harbour-McMenamin D, Smith EM, Blalock JE: Production of immunoreactive chorionic gonadotropin during mixed lymphocyte reactions: A possible mechanism for genetic diversity. Proc Natl Acad Sci USA 1986;83:6834–6838.
64 Fuchs T, Hammarstrom L, Smith CI, Brundin J: Sex-dependent induction of human suppressor T cells by chorionic gonadotropin. J Reprod Immunol 1982;4:185–190.
65 Ricketts RM, Jones DB: Differential effect of human chorionic gonadotropin on lymphocyte proliferation induced by mitogens. J Reprod Immunol 1985;7:225–232.

66 Cooper JR, Bloom FE, Roth RH: The Biochemical Basis of Neuropharmacology, ed 5. New York, Oxford University Press, 1986, pp 352–393.
67 Pernow B: Substance P. Pharmacol Rev 1983;35:85–141.
68 Payan DG, Brewster DR, Goetzl EJ: Stereospecific receptors for substance P on cultured human IM-9 lymphoblasts. J Immunol 1984;133:3260–3265.
69 Payan DG, McGillis JP, Organist ML: Binding characteristics and affinity labeling of protein constituents of the human IM-9 lymphoblast receptor for substance P. J Biol Chem 1986;261:14321–14329.
70 McGillis JP, Mitsuhashi M, Payan DG: Immunomodulation by tachykinin neuropeptides. Ann NY Acad Sci 1990;594:85–95.
71 Organist ML, Harvey J, McGillis JP, Mitsuhashi M, Melera P, Payan DG: Characterization of a monoclonal antibody against the lymphoblast substance P receptor. J Immunol 1987;139:001–005.
72 Foreman J, Jordan C: Histamine release and vascular changes induced by neuropeptides. Agents Actions 1983;13:105–116.
73 Goetzl EJ, Chernov T, Renold F, Payan DG: Neuropeptide regulation of the expression of immediate hypersensitivity. J Immunol 1985;135:802s–805s.
74 Marasco WA, Showell HJ, Becker EL: Substance P binds to the formylpeptide chemotaxis receptor on the rabbit neutrophil. Biochem Biophys Res Commun 1982; 99:1065–1072.
75 Bar-shavit Z, Goldman R, Stabinsky R, Gottlieb P, Fridkin M, Teichberg VI, Blumberg S: Enhancement of phagocytosis – A newly found activity of substance P in its N-terminal tetrapeptide sequence. Biochem Biophys Res Commun 1980;94: 1445–1451.
76 Hartung H-P, Wolters K, Toyka KV: Substance P: Binding properties and studies on cellular responses in guinea pig macrophages. J Immunol 1986;136:3856–3863.
77 Payan DG, Brewster DR, Goetzl EJ: Specific stimulation of human T lymphocytes by substance P. J Immunol 1983;13:1613–1615.
78 Stanisz AM, Befus D, Bienenstock J: Differential effects of vasoactive intestinal peptide, substance P, and somatostatin on immunoglobulin synthesis and proliferation by lymphocytes from Peyer's patches, mesenteric lymph nodes, and spleen. J Immunol 1986;136:152–156.
79 Payan DG, Brewster DR, Missirian-Bastian A, Goetzl EJ: Substance P recognition by a subset of human T lymphocytes. J Clin Invest 1984;74:1532–1539.
80 Ruff MR, Wahl SM, Pert CB: Substance P receptor-mediated chemotaxis of human monocytes. Peptides 1985;2(suppl 6):107–111.
81 Lotz M, Vaughan JH, Carson DA: Effects of neuropeptides on production of inflammatory cytokines by human monocytes. Science 1988;241:1218–1221.
82 Lotz M, Carson DA, Vaughan JH: Substance P activation of rheumatoid synoviocytes: Neural pathway in pathogenesis of arthritis. Science 1987;235:893–895.
83 Yankner BA, Duffy LK, Kirschner DA: Neurotrophic and neurotoxic effects of amyloid β protein: Reversal by tachykinin neuropeptides. Science 1990;250:279–282.
84 Mitsuhashi M, Akitaya T, Payan DG: Amyloid β substituent peptides do not interact with the substance P receptor expressed in cultured cells. Mol Brain Res, submitted.
85 Costa M, Furness JB: The origins, pathways, and terminations of neurons with VIP-like immunoreactivity in the guinea pig small intestine. Neuroscience 1983;8:665–676.

86 Morel G, Besson J, Rosselin G, Dubois PM: Ultrastructural evidence for endogenous vasoactive intestinal peptide-like immunoreactivity in the pituitary gland. Neuroendocrinology 1982;34:85–89.
87 Weihe E, Reinecke M: Peptidergic innervation of the mammalian sinus nodes: Vasoactive intestinal polypeptide, neurotensin, substance P. Neurosci Lett 1981;26: 283–288.
88 Barajas L, Sokolski DN, Lechago J: Vasoactive intestinal polypeptide-immunoreactive nerves in the kidney. Neurosci Lett 1983;43:263–269.
89 Bloom SR, Polak JM: Regulatory peptides and the skin. Clin Exp Dermatol 1983;8: 3–18.
90 Felten DL, Felten SY, Carlson SL, Olschowka JA, Livnat S: Noradrenergic and peptidergic innervation of lymphoid tissue. J Immunol 1985;135:755s–765s.
91 Lundberg JM, Pernow AA, Hokfelt T: Neuropeptide Y-, substance P- and VIP-immunoreactive nerves in cat spleen in relation to autonomic vascular and volume control. Cell Tissue Res 1985;239:9–18.
92 O'Dorisio MS, Wood CL, O'Dorisio TM: Vasoactive intestinal peptide and neuropeptide modulation of the immune response. J Immunol 1985;135:792s–799s.
93 Frawley LS, Neill JD: Stimulation of prolactin secretion in rhesus monkeys by vasoactive intestinal polypeptide. Neuroendocrinology 1981;33:79–83.
94 Feliu JE, Mojena M, Silvestre RA, Monge J, Marco J: Stimulatory effect of vasoactive intestinal peptide on glycogenolysis and gluconeogenesis in isolated rat hepatocytes: Antagonism by insulin. Endocrinology 1983;112:2120–2127.
95 Magistretti PJ, Morrison JH, Shoemaker WJ, Sapin V, Bloom FE: Vasoactive intestinal polypeptide induces glycogenolysis in mouse control of energy metabolism. Proc Natl Acad Sci USA 1981;78:6535–6539.
96 Epelbaum J, Tapia-Arancibia L, Besson J, Rotszteyn WH, Kordon C: Vasoactive intestinal peptide inhibits release of somatostatin from hypothalamus in vitro. Eur J Pharmacol 1979;58:493–495.
97 Said SI, Geumi A, Hara N: Bronchodilator effect of VIP in vivo: Protection against bronchoconstriction induced by histamine or prostaglandin $F_{2\alpha}$; in Said (ed): Vasoactive Intestinal Peptide. New York, Raven Press, 1982, pp 185–192.
98 Ottesen B, Gerstenberg T, Ulrichsen H, Manthrope T, Fahrenkrug J, Wagner G: Vasoactive intestinal polypeptide (VIP) increases vaginal blood flow and inhibits uterine smooth muscle activity in women. Eur J Clin Invest 1983;13:321–324.
99 O'Dorisio MS, O'Dorisio TM, Cataland S, Balcerzak SP: VIP as a biochemical marker for polymorphonuclear leukocytes. J Lab Clin Med 1980;96:666–672.
100 Cutz E, Chan W, Track NS, Goth A, Said S: Release of vasoactive intestinal polypeptide in mast cells by histamine liberators. Nature (Lond) 1978;275:661–662.
101 Goetzl EJ, Sreedharan SP, Turck CW: Structurally distinctive vasoactive intestinal peptides from rat basophilic leukemia cells. J Biol Chem 1988;263:9083–9086.
102 Ottaway CA, Greenberg GR: Interaction of vasoactive intestinal peptide with mouse lymphocytes: Specific binding and the modulation of mitogen responses. J Immunol 1984;132:417–423.
103 Ottaway CA, Bernaerts C, Chan B, Greenberg GR: Specific binding of VIP to human circulating mononuclear cells. Can J Physiol Pharmacol 1983;61:664–671.
104 Ottaway CA: Selective effects of vasoactive intestinal peptide on the mitogenic response of murine T cells. Immunology 1987;62:291–297.

105 Elson CO, Heck JA, Strober W: T-cell regulation of murine IgA synthesis. J Exp Med 1979;149:632–643.
106 Richman LK, Graeff AS, Yarchoan R, Strober W: Simultaneous induction of antigen-specific IgA helper T cell and IgG suppressor T cells in the murine Peyer's patch after protein feeding. J Immunol 1981;126:2079–2083.
107 Rola-Pleszczynski M, Bolduc D, St Pierre S: The effects of vasoactive intestinal peptide on human natural killer cell function. J Immunol 1985;135:2569–2573.
108 Holst JJ, Fahrenkrug J, Jensen SL, Nielsen OU: Peptidergic innervation of the pancreas; in Bloom, Polak (eds): Gut Hormones. Edinburgh, Churchill Livingstone, 1981, pp 495–511.
109 Koff WC, Dunegan MA: Modulation of macrophage-mediated tumoricidal activity by neuropeptides and neurohormones. J Immunol 1985;135:350–354.
110 Yiangou Y, Serrano R, Bloom SR, Peña J, Festenstein H: Effects of prepro-vasoactive intestinal peptide-derived peptides on the murine immune response. J Neuroimmunol 1990;29:65–72.
111 Moore TC: Modification of lymphocyte traffic by vasoactive neurotransmitter substances. Immunology 1984;52:511–518.
112 Ottaway CA: In vitro alteration of receptors for vasoactive intestinal peptide changes in vivo localization of mouse T cells. J Exp Med 1984;160:1054–1069.
113 O'Dorisio MS, Hermina NS, O'Dorisio TM, Balcerzak SP: Vasoactive intestinal polypeptide modulation of lymphocyte adenylate cyclase. J Immunol 1981;127: 2551–2554.
114 Beed EA, O'Dorisio MS, O'Dorisio TM, Gaginella TS: Demonstration of a functional receptor for vasoactive intestinal polypeptide on Molt-4B lymphoblasts. Regul Pept 1983;6:1–12.
115 O'Dorisio MS: Biochemical characteristics of receptors for vasoactive intestinal polypeptide in nervous, endocrine, and immune systems. Fed Proc 1987;46:192–195.
116 Ottaway CA, Lay TE, Greenberg GR: High affinity specific binding of vasoactive intestinal peptide to human circulating T cells, B cells and large granular lymphocytes. J Neuroimmunol 1990;29:149–155.
117 O'Dorisio MS, Wood CL, Wenger GD, Vassallo LM: Cyclic AMP-dependent protein kinase in Molt-4B lymphoblasts: Identification by photoaffinity labeling and activation in intact cells by the neuropeptides vasoactive intestinal polypeptide and peptide histidine isoleucine. J Immunol 1985;134:4078–4086.
118 Gilman AG: G proteins and dual control of adenylate cyclase. Cell 1984;36:577–579.
119 Laycock JF, Wise PH: The hypothalamo-hypophyseal system; in Essential Endocrinology, ed 2. New York, Oxford University Press, 1983, pp 29–76.
120 Rudman D: Growth hormone, body composition, and aging. J Am Geriatr Soc 1985; 33:800–807.
121 Rudman D, et al: Effects of human growth hormone in men over 60 years old. N Engl J Med 1990;323:1–6.
122 Weigent DA, Baxter JB, Wear WE, Smith LR, Bost KL, Blalock JE: Production of immunoreactive growth hormone by mononuclear leukocytes. FASEB J 1988;2: 2812–2818.
123 Hattori N, Shimatsu A, Sugita M, Kumagi S, Imura H: Immunoreactive growth hormone (GH) secretion by human lymphocytes: augmented release by exogenous GH. Biochem Biophys Res Commun 1990;168:396–401.

124 Hiestand PC, Mekler P, Nordmann R, Grieder A, Permmongkol C: Prolactin as a modulator of lymphocyte responsiveness provides a possible mechanism of action for cyclosporine. Proc Natl Acad Sci USA 1986;83:2599–2603.
125 Smith PE: The effect of hypophysectomy upon the involution of the thymus in the rat. Anat Rec 1930;47:119–129.
126 Berczi I, Nagy E: A possible role of prolactin in adjuvant arthritis. Arthritis Rheum 1982;25:591–594.
127 Nagy E, Berczi I, Friesen HG: Regulation of immunity in rats by lactogenic and growth hormones. Acta Endocrinol (Copenh) 1983;102:351–357.
128 Arranbrecht S: Specific binding of growth hormone to thymocytes. Nature (Lond) 1974;252:255–257.
129 Lesniak MA, Gorden P, Roth J, Gavin JR III: Binding of ^{125}I-human growth hormone to specific receptors in human cultured lymphocytes. J Biol Chem 1974; 249:1661–1667.
130 Kiess W, Butenandt O: Specific growth hormone receptors on human peripheral mononuclear cells. Reexpression, identification, and characterization. J Clin Endocrinol Metab 1985;60:740–746.
131 Russell DH, Kibler R, Matrisian L, Larson DF, Poulos B, Magun BE: Prolactin receptors on human T and B lymphocytes. Antagonism of prolactin binding by cyclosporine. J Immunol 1985;134:3027–3031.
132 Russell DH, Matrisian L, Kibler R, Larson DF, Poulos B, Magun BD: Prolactin receptors on human lymphocytes and their modulation by cyclosporine. Biochem Biophys Res Commun 1984;121:899–906.
133 Kelley KW, Brief S, Westly HJ, Novakofski J, Bechtel PJ, Simon J, Walker EB: GH_3 pituitary adenoma cells can reverse thymic aging rats. Proc Natl Acad Sci USA 1986; 83:5663–5667.
134 Prysor-Jones RA, Jenkins JS: Effect of excessive secretion of growth hormone on tissues of the rat, with particular reference to the heart and skeletal muscle. J Endocrinol 1980;85:75–82.
135 Seo H, Refetoff S, Fang VS: Induction of hypothyroidism and hypoprolactinemia by growth hormone producing rat pituitary tumors. Endocrinology 1977;100:216–226.
136 Davila DR, Brief S, Simon J, Hammer RE, Brinster RL, Kelley KW: Role of growth hormone in regulating T-dependent immune events in aged, nude, and transgenic rodents. J Neurosci Res 1987;18:108–116.
137 Kelley KW: The role of growth hormone in modulation of the immune response. Ann NY Acad Sci 1990;594:95–103.
138 Weigent DA, Blalock JE, LeBoeuf RD: An antisense oligodeoxynucleotide to growth hormone messenger ribonucleic acid inhibits lymphocyte proliferation. Endocrinology 1991;128:2053–2057.
139 Nicoll CS, Bern HA: On the actions of prolactin among the vertebrates: Is there a common denominator? In Wolstenholme GEW, Knight J (eds): Lactogenic Hormones. London, Churchill Livingstone, 1972, pp 299–317.
140 Bernton EW, Meltzer MS, Holaday JW: Suppression of macrophage activation and T-lymphocyte function in hypoprolactinemic mice. Science 1988;239:401–404.
141 Wilner ML, Ettenger RB, Koyle MA, Rosenthal JT: The effect of hypoprolactinemia alone and in combination with cyclosporine on allograft rejection. Transplantation 1990;49:264–267.

142 Carrier M, Russell DH, Wild JC, Emery RW, Copeland JG: Prolactin as a marker of rejection in human heart transplantation. J Heart Transplant 1987;6:290–292.
143 Berczi I, Nagy E, DeToledo SM, Matusik RJ, Friesen HG: Pituitary hormones regulate c-myc and DNA synthesis in lymphoid tissue. J Immunol 1991;146:2201–2206.
144 Russell DH: New aspects of prolactin and immunity: a lymphocyte-derived prolactin-like product and nuclear protein kinase C activation. TiPS 1989;10:40–44.
145 Tabor CW, Tabor H: Polyamines. Annu Rev Biochem 1984;53:749–790.
146 Clevenger CV, Russell DH, Appasamy PM, Prystowsky MB: Regulation in interleukin-2-driven T-lymphocyte proliferation by prolactin. Proc Natl Acad Sci USA 1990;87:6460–6464.
147 Turkington RW: Ectopic production of prolactin. N Engl J Med 1971;285:1455–1458.
148 Rosen SW, Weintraub BD, Aaronson SA: Nonrandom ectopic protein production by malignant cells: direct evidence in vitro. J Clin Endocrinol Metab 1980;50:834–841.
149 DiMattia GE, Gellersen B, Bohnet HG, Friesen HG: A human B-lymphoblastoid cell line produces prolactin. Endocrinology 1988;122:2508–2517.
150 Montgomery DW, Zukoski CF, Shah GN, Buckley AR, Pacholczyk T, Russell DH: Concanavalin A-stimulated murine splenocytes produce a factor with prolactin-like bioactivity and immunoreactivity. Biochem Biophys Res Commun 1987;145:692–698.
151 Schuller LA, Hurley WL: Molecular cloning of a prolactin-related mRNA expressed in bovine placenta. Proc Natl Acad Sci USA 1987;84:5650–5654.
152 Duckworth ML, Kirk KL, Friesen HG: Isolation and identification of a cDNA clone of rat placental lactogen. II. J Biol Chem 1986;261:10871–10878.
153 Linzer DIH, Lee S-J, Ogren L, Talamantes F, Nathans D: Identification of proliferin mRNA and protein in mouse placenta. Proc Natl Acad Sci USA 1985;82:4356–4359.
154 Payan DG, Hess CA, Goetzl EJ: Inhibition by somatostatin of the proliferation of T-lymphocytes and Molt-4 lymphoblasts. Cell Immunol 1984;84:433–438.
155 Malec P, Zeman K, Markiewicz K, Tchorzewski H, Nowak Z, Baj Z: Short-term somatostatin infusion affects T lymphocyte responsiveness in humans. Immunopharmacology 1989;17:45–49.
156 Muscettola M, Grasso G: Somatostatin and vasoactive intestinal peptide reduce interferon-gamma production by human peripheral blood mononuclear cells. Immunobiology 1990;180:419–430.
157 Wagner M, Hengst K, Zierden E, Gerlach U: Investigations of the antiproliferative effect of somatostatin in man and rats. Metab Clin Exp 1979;27:1381–1386.
158 Hinterberger W, Cerny C, Kinast M, Pointer H, Trag KM: Somatostatin reduces the release of colony-stimulating activity (CSA) from PHA-activated mouse spleen lymphocytes. Experientia 1977;34:860–862.
159 Sreedharan SP, Kodama KT, Peterson KE, Goetzl EJ: Distinct subsets of somatostatin receptors on cultured human lymphocytes. J Biol Chem 1989;264:949–952.
160 Nakamura H, Koike T, Hiruma K, Sate T, Tomioka H, Yoshida S: Identification of lymphoid cell lines bearing receptors for somatostatin. Immunology 1987;62:655–658.
161 Pawlikowski M, Stepien M, Kunert-Radek J, Zelazowski P, Schally AV: Immunomodulatory action of somatostatin. Ann NY Acad Sci 1987;496:233–239.
162 Kirk CJ, Creba JA, Downes CP, Michell RH: Hormone-stimulated metabolism of inositol lipids and its relationship to hepatic receptor function. Biochem Soc Trans 1981;9:377–379.

163 Rhodes D, Prpic V, Exton JH, Blackmore PF: Stimulation of phosphate hydrolysis in hepatocytes by vasopressin. J Biol Chem 1983;258:2770–2773.
164 Thomas AP, Marks JS, Coll KE, Williamson JR: Quantitation and early kinetics of inositol lipid changes induced by vasopressin in isolated and cultured hepatocytes. J Biol Chem 1983;258:5716–5725.
165 Berridge MJ: Inositol trisphosphate and diacylglycerol: Two interacting second messengers. Annu Rev Biochem 1987;56:159–193.
166 Hirasawa K, Nishizuka Y: Phosphatidylinositol turnover in receptor mechanism and signal transduction. Annu Rev Pharmacol Toxicol 1985;25:147–170.
167 Ganz MB, Boyarsky G, Sterzel RB, Boron WF: Arginine vasopressin enhances pH_i regulation in the presence of HCO_3^- by stimulating three acid-base transport systems. Nature 1989;337:648–651.
168 Irvine RF: How is the level of free arachidonic acid controlled in mammalian cells? Biochem J 1982;204:3–16.
169 McPhail LC, Clayton CC, Snyderman R: A potential second messenger role for unsaturated fatty acids: Activation of calcium-dependent protein kinase. Science 1984;224:622–625.
170 Hadden JW, Coffey RG: Cyclic nucleotides in mitogen-induced lymphocyte proliferation. Immunol Today 1982;3:299–304.
171 Johnson HM, Torres BA: Leukotrienes: Positive signals for regulation of gamma interferon production. J Immunol 1984;132:413–416.
172 Johnson HM, Russell JK, Torres BA: Second messenger role of arachidonic acid and its metabolites in gamma interferon production. J Immunol 1986;137:3053–3056.
173 Johnson HM, Archer DL, Torres BA: Cyclic GMP as the second messenger in helper cell requirement for gamma interferon production. J Immunol 1982;129:2570–2572.
174 Johnson HM, Vassallo T, Torres BA: Interleukin-2-mediated events in gamma interferon production are calcium dependent at more than one site. J Immunol 1985; 134:967–970.
175 Gomperts BD, Barrowman MM, Cockcroft S: Dual role for guanine nucleotides in stimulus-secretion coupling. Fed Proc 1986;45:2156–2161.

Howard M. Johnson, PhD, Department of Microbiology and Cell Science,
The University of Florida, Gainesville, FL 32611 (USA)

Blalock JE (ed): Neuroimmunoendocrinology, 2nd rev ed.
Chem Immunol. Basel, Karger, 1992, vol 52, pp 84–105

Neuroendocrine Peptide Receptors on Cells of the Immune System

Daniel J.J. Carr

Department of Physiology and Biophysics, University of Alabama at Birmingham, Ala., USA

Introduction

The interaction between the immune and neuroendocrine systems takes place, in part, through endocrine and paracrine routes by neurotransmitters and peptide hormones. Neurotransmitters are secreted by sympathetic fibers innervating immune organs [Felten et al., this volume] while certain neuropeptides and hormones are secreted through the hypophyseal portal system. These signal molecules comprise a diverse family of chemical structures which interact with immune cell-surface receptors. These interactions ultimately result in some modification in the functionality of the target. The fact that cells of the immune system display various neuropeptide receptors and produce the neuropeptide molecules to these receptors adds another dimension to the neuroendocrine-immune axis: autocrine or paracrine regulation of an immune response by neuropeptides. Another important feature of this system is that circulating peripheral immunocytes establish a mobile source of neuropeptides which locally may achieve physiologic or pharmacologic levels. The above discussion serves to illustrate how the immune system is an important and often neglected 'arm' of the endocrine system that should be considered before a more complete model of endocrine physiology can be fully understood.

In defining neuropeptide receptors on immunocytes, three criteria have to be established: (1) identify the receptor on the basis of pharmacological and radioreceptor assays which determine affinity, ligand selectivity, and antagonists of the receptor; (2) structurally characterize the receptor and identify intracellular second messenger signaling pathways utilized following ligand-receptor interaction, and (3) establish the biological consequences of receptor activation relative to immunocompetence. In some instances, immunocyte neuropeptide receptor characterizations fall short of fulfilling all the above criteria primarily due to (1) the low number of receptors which exists on the membrane surface of the cell; (2) the lack of ligand probes which can be employed to structurally define the receptor binding site, and (3) the absence of pure antagonists which can be used in a biological setting.

Thus, the intent of this review is to broadly focus on the identification of neuropeptide receptors on cells of the immune system using the above criteria where applicable.

Arginine Vasopressin Receptors

Arginine vasopressin (AVP) is normally associated with the regulation of antidiuresis [1] and glycogenolysis [2]. However, it has more recently been implicated as an immune regulator. AVP receptors have been categorized into V_1 receptors (occurring in hepatic cells) which cause calcium mobilization [3] or V_2 receptors (antidiuretic receptor) which stimulate cAMP production [4]. Cells of the immune system possess V_1 or V_1-like receptors. Functionally, AVP (10^{-9}–10^{-10} *M*) has been shown to facilitate the production of γ-interferon (γ-IFN) by substituting for interleukin-2 (IL-2) as a helper signal in murine splenocytes [5]. This response is blocked by the V_1 antagonist, [d(CH_2)1/5 Tyr-(Me)]AVP but not by another V_1 antagonist, [d(CH_2)1/5D Tyr(Et(^{2}Vai^4]AVP or V_2 antagonists [6] indiating a novel V_1 receptor binding domain on lymphocytes. These biological observations correlate well with radioreceptor studies recently reported with AVP and lymphocytes [7]. Human peripheral blood lymphocytes (PBMCs) possess a saturable, high affinity ($K_d = 0.47 \pm 0.17$ n*M*) selective cell surface receptor.

The results from these studies also show the receptor density is significantly higher on PBMCs taken from female donors versus males, although no correlation is found between the number of receptor sites and the menstrual cycle. Additional findings also indicate a ligand selectivity of an AVP binding peptide that was species-dependent [8]. Aside from the possible variants of the V_1 receptor on immune cells, the AVP receptor on immunocytes apparently is functionally identical to those found on pituitary cells with regard to corticotropin-releasing hormone (CRH). Specifically, AVP similarly enhances CRH activity on pituitary cells [9] and lymphocytes (see section on lymphocyte CRH receptors).

Corticotropin Receptor

Adrenocorticotropic hormone (ACTH), a 39-amino acid peptide, is an important immunoregulatory molecule especially in humoral immunity. Initial studies indicated ACTH (10^{-6}–10^{-7} *M*) suppressed T-dependent and T-independent antibody production in vitro [10]. Unlike $ACTH_{1-39}$, $ACTH_{1-24}$ did not affect antibody production [10] although it has been found to modulate (enhance or suppress depending on the donor) lymphocyte proliferation [11]. ACTH has also been implicated as a growth and differentiation factor [12] which may work in a paracrine or autocrine [13] fashion. In fact, recent evidence indicates ACTH increases LPS-activated IgM production by the murine B cell line CH12.LX.C4.5F5 at low concentrations (10^{-9}–10^{-13} *M*) whereas high concentrations (10^{-6} *M*) reduce IgM production [14].

The effects of ACTH on immunocompetence occur through cell surface ACTH receptors. Rat lymphocytes possess both high (K_d = 88 p*M*; B_{max} = 1,100 sites/cell) and low (K_d = 4.2 n*M*; B_{max} = 30,000 sites/cell) affinity receptor binding sites while unstimulated thymocytes display few receptors [15]. In addition, B lymphocytes possess 3 times the number of high affinity binding sites compared to T lymphocytes [15]. The number of ACTH receptors present on lymphocytes and thymocytes is dependent on whether the cells are in a 'resting' or 'activated' state. For example, treatment of lymphocytes or thymocytes with the mitogen concanavalin A (ConA) increases the number of high affinity ACTH receptors on B and T lymphocytes by 2- to 3-fold. ConA-stimulated thymocytes result in a 100-fold increase in the high affinity ACTH receptor binding site [15]. ACTH binding to its receptor initiates a signal transduction pathway that involves cyclic AMP mobilization of CA^{2+}.

Specifically, ACTH (10^{-5}–10^{-10} *M*) enhances the production of cAMP by murine splenocytes in a dose-dependent fashion as early as 5 min [16]. The maximum effect is achieved at 10^{-8} *M* ACTH. Additionally, stimulation of cAMP synthesis is directly correlated with ACTH receptor expression since lymphocyte cAMP production by immune cells parallels receptor expression [16]. Physiologic levels of ACTH (10^{-9}–10^{-12} *M*) also increase Ca^{2+} influx in rat lymphocytes with maximum intracellular Ca^{2+} levels occurring at 1 n*M* ACTH [Clarke and Blalock, submitted]. These effects are most likely due to the activation of the high affinity receptor since the biological responses are observed at 10^{-10}–10^{-12} *M* ACTH concentrations. Furthermore, it has been suggested that ACTH binding to this receptor involves, either directly of indirectly, the activation of a K^+-dependent calcium channel resulting in an influx of Ca^{2+} and subsequently proliferation of lymphocytes [Ben Clarke, pers. commun.].

The effects of ACTH binding to the low affinity receptor on lymphocytes on cellular responses are not as well understood as the effects of ACTH binding to the high affinity receptor. It may be that ACTH binding to the low affinity receptor affects regulatory elements in the adenylate cyclase system. This hypothesis is similar to the one advanced for ACTH stimulation of steroidogenesis in adrenal cells [17]. Although this idea remains unconfirmed, we do know that at concentrations of ACTH ($>10^{-9}$ *M*) in which cAMP synthesis is nearly maximal in lymphocytes [16], immunoglobulin synthesis is suppressed, implicating the low affinity receptor in this regulatory circuit [10].

β-Endorphin Receptors

β-Endorphin-specific receptors were first identified on cultured human lymphocytes using ^{125}I-β[*D*-Ala^2]endorphin [18]. This receptor is specific for β-endorphin since α- or γ-endorphins, enkephalins, morphine, and naloxone are ineffective at displacing radiolabeled β-endorphin. Unlabeled β-endorphin can inhibit binding of ^{125}I-β-endorphin to lymphocytes with an $IC_{50} \sim 50$ n*M*. From these early studies stemmed a host of functional studies identifying β-endorphin-selective pathways that suppressed or enhanced a variety of immune parameters including lymphocyte proliferation, IL-2 production, and γ-IFN production [19]. Some of these effects are naloxone-irreversible indicating a nonopioid component of the peptide is required to achieve some effects. When the N-terminal portion of β-endorphin (required for interactions with opioid receptors) was intact, the molecule showed a greater degree of biological activity as measured by lymphocyte proliferation compared to truncated forms of β-endorphin, i.e. β-endorphin$_{6-31}$, β-endorphin$_{18-21}$ [20]. Biochemical studies on the β-endorphin-specific receptors on the murine EL-4 thymoma [21] have provided information regarding the molecular structure of the receptor. Using chemical cross-linking techniques and polyacrylamide gel electrophoresis (PAGE), the β-endorphin binding site is a polypeptide with a molecular weight of 72 kDa under reducing and nonreducing conditions [22]. Other studies using murine leukocytes show they also display β-endorphin-specific receptors ($K_d = 4$ n*M*; $B_{max} = 0.2$ fmol/10^6 cells) only after the culturing of these cells for 96 h [23]. Chemical cross-linking with ^{125}I-β-endorphin and PAGE analysis of the receptor on these cells shows that three polypeptides with molecular weights of 66, 57, and 44 kDa are specifically labeled [23, 24]. Recent studies suggest the 66-kDa protein is most likely a serum contaminant such as albumin [25]. β-Endorphin-specific receptors have also been found on normal human and Epstein-Barr virus-transformed B lymphocytes [26]. These receptors reportedly regulate Ca^{2+} uptake in rat thymocytes [27].

The properties of the β-endorphin-specific receptor on lymphocytes are remarkably similar to the properties of neuronal opioid receptors [28]. Specifically, they have similar molecular weight binding polypeptides and both are apparently coupled to G proteins [24, 28]. They differ in that lymphocyte β-endorphin-specific receptors are insensitive to naloxone while neuronal opioid receptors are naloxone-sensitive. The similarities between neuronal opioid receptors and lymphocyte β-endorphin-specific receptors seem to indicate a common element (structural) may be shared between opioid and β-endorphin-specific receptors. In support of this hypothesis, a previous study using chemical cross-linking of ^{125}I-β-endorphin to murine lymphocytes showed the labeling of a 46-kDa protein as determined by PAGE [29]. In addition, the labeling was opioid-sensitive since in the presence of excess, unlabeled naloxone, no 46-kDa protein was apparent. Likewise, a 46-kDa protein has also been identified as a common structural element shared by the neural K, μ, and δ opioid receptor classes [30]. The observations showing a 44-kDa protein associated with the lymphocyte β-endorphin-specific receptor [23] coupled with the data showing a 46-kDa polypeptide associated with neuronal [30] and lymphoid opioid receptors [29] suggests the 44- to 46-kDa protein moiety may be a shared entity among these receptors.

Growth Hormone Receptors

Growth hormone (GH) has been proposed to be an important immunoregulatory molecule since 1959 [31]. Confirmation of this suggestion has occurred through a number of studies both in vitro and in vivo [32]. The effects of GH on lymphocyte functional responses are mediated through specific cell surface receptors which have been identified on both thymocytes [33] and lymphocytes [34–36]. The lymphocyte GH receptor has a K_d = 1–10 n*M* with 6,500–8,000 binding sites/cell. GH may affect immunological responses in an autocrine or paracrine manner since lymphocytes and macrophages produce GH [37, 38]. Lymphocyte production of GH is approximately 15 pg/10^6 cells with B lymphocytes synthesizing more than T helper (T_H) cells [38]. T suppressor (Ts) cells appear to produce little if any GH. The lymphocyte/macrophage-derived GH has been shown to bind to the lymphocyte GH receptor [39] indicating that it may serve in an autocrine or paracrine manner in vivo. Lymphocyte GH production is strongly implicated in the maintenance of immunocompetence since treatment of hypophysectomized with GH restores immune function [40]. At present, no intracellular signal transduction pathways have been conclusively identified in immune cells following GH ligand-receptor interactions. Future studies using the recently cloned cDNA for the rabbit liver GH receptor [41] should provide information on signal processing after GH binding to its receptor.

Nerve Growth Factor Receptors

Nerve growth factor (NGF) was originally thought to be associated with the development and differentiation of peripheral sympathetic and sensory neurons [42]. However, the results of additional studies indicate NGF affects other nonneuronal tissues including cells of the immune system. For example, NGF stimulates rat mast cell growth, enhances rat lymphocyte proliferation, and increases both IL-2 receptor expression and IL-2 production by cultured lymphocytes [43,44]. Likewise, NGF significantly enhances human B lymphocyte proliferation and augments IgM but not IgG, IgA, or IgE production by cultured cells [45]. The NGF receptor is most abundant on B lymphocytes (3.94 ± 0.08 fmol/10^8 cells) followed by T lymphocytes (2.76 ± 0.3 fmol/10^8 cells) and monocytes (1.06 ± 0.19 fmol/10^8 cells) with an affinity in the nanomolar to subnanomolar range [45, 46].

The molecular structure of the lymphocyte NGF receptor has been studied by chemical cross-linking and PAGE. The results using ^{125}I-NGF show the labeling of a single polypeptide with a molecular weight of 85 kDa [45]. At this time, the molecular mechanisms regulating lymphocyte NGF receptor expression as well as intracellular signaling pathways coupled to the receptor are not known. However, both IL-1 and IL-6 have been implicated in eliciting NGF production in cerebral and peripheral tissues which suggests NGF may be a likely candidate as a local B cell growth factor in brain tissue following head trauma [45].

Opioid Receptors

Opioid receptors have been categorized into predominately three classes including μ, δ, and κ. Early investigations indicated both μ- and δ-class opioid receptors were expressed on T cells using T lymphocytes in functional assays [47, 48]. These initial studies were substantiated using dihydromorphine, naloxone, and [leu]-enkephalin in radioreceptor assays [10]. A saturable, high affinity binding site (K_d = 8–10 n*M*, 300–4,000 sites/cell) was reported on granulocytes and monocytes using [^{3}H]dihydromorphine [49]. The immunocyte opioid receptors were also shown to be closely associated with lipid moieties similar to what one observes with brain opioid receptors [50]. More recently, a single saturable binding site for [^{3}H]bremazocine with a K_d of 60 n*M* (B_{max} = 2.7 pmol/10^6 cells) has been identified on the EL-4 thymoma cell line [51]. Although the binding site does not exhibit stereoselectivity (a hallmark for neuronal opioid receptors), it is class-selective. Likewise, a high affinity (K_d = 17 n*M*; B_{max} = 54 fmol/10^6 cells) enantioselective κ-like binding site has been identified on the macrophage cell line, P388d_1 [52]. The macrophage site increases Ca^{2+} mobilization in a time-dependent manner [53]. Similarly, a saturable, naloxone-selective binding site has also been identified on rat lymphocyte membranes (K_d = 960 n*M*; B_{max} = 5 n*M*/2 × 10^6 cells) which is sensitive to morphine (IC_{50} = 130 n*M*) but not [leu]-enkephalin, [met]-enkephalin, or β-endorphin [54].

The μ- and δ-class opioid receptor binding sites appear to be structurally related. Specifically, using [^{3}H]cis-(+)-3-methylfentanyl-isothiocyanate (δ-class-selective ligand), a binding site with a molecular weight of 58 kDa is specifially labeled on murine lymphocytes [55] and the P388d_1 macrophage cell line [52]. The μ-selective, site-directed acylating agent, [^{3}H]2–(*p*-ethoxybenzyl)-1-[N,N-diethylamino]ethyl-5-isothiocyanato-benzimidazole labels a lymphocyte membrane protein exhibiting μ-class selectivity which also has a molecular weight of 58 kDa [56]. Although the μ and δ opioid receptors have structurally related binding sites, κ-receptors are unrelated. Specifically, the κ-selective, site-directed acylating agent, [^{3}H](1s,2s)-(–)-trans-2-isothiocyanato-N-methyl-N-[2-(1-pyrrolidinyl)cyclohexyl]benzeneacetamide, labels a protein on lymphocytes exhibiting κ-class selectivity which has a molecular weight of 38–42 kDa [57].

Opioid receptor activation leads to the stimulation of guanylate cyclase [58], inhibition of adenylate cyclase [58, 59], reduction in potassium conductance [29, 60], and enhancement in Ca^{2+} uptake [53, 60]. The fact that opioids modulate a host of immune responses in vitro [19] and in vivo

[61–63] indicates the relevance of the receptors on cells of the immune system. In addition, the observation that opioid peptides are produced by immunocytes [see Blalock, this volume] and are active antinociceptive agents [64] illustrate the remarkably interrelatedness between the immune and neuroendocrine systems based on this opioid afferent/efferent pathway.

Prolactin Receptor

Similar to GH, prolactin is apparently necessary for the maintenance and responsiveness of the immune system. For example, hypophysectomized animals which have an impaired immune response become fully immunocompetent following the exogenous administration of prolactin [40]. The results of in vitro studies have shown the addition of antiprolactin antibodies to culture media reduces mitogen-stimulated proliferation of T and B lymphocytes [65]. The results of more detailed studies also show that prolactin stimulates the expression of genes associated with lymphocyte activation and proliferation. For example, prolactin induces a novel growth response gene termed c25, C-myc, and Nb29 (a heat shock protein 70 homologue) in NB2 T cells [66]. Although prolactin appears to be a positive regulator of immune function, high levels of serum prolactin are associated with the impairment of natural killer cell function and absolute number [67]. Therefore, prolactin may be a positive or negative immunomodulator depending on the target cell. Saturable, high affinity (K_d = 1.7 n*M*; B_{max} = 360 sites/cell) prolactin receptors are found on human lymphocytes [68]. Similarly, large granular lymphocytes (LGLs) which mediate the majority of natural killer activity possess saturable, high affinity (K_d = 300 p*M*; B_{max} = 660 sites/cell) binding sites [69]. In both LGLs and B and T lymphocytes, the immunosuppressive cyclic endecapeptide ciclosporin can regulate receptor expression. High concentrations of ciclosporin (10^{-9}–10^{-10} *M*) decrease prolactin receptor expression while low concentrations (10^{-9}–10^{-10} *M*) increase the number of cell surface prolactin receptors [69, 70]. Since antibodies to prolactin have been shown to neutralize CTLL-2 proliferation to IL-2 [71] and prolactin induces IL-2 receptor expression on lymphocytes [72], some of the immunosuppressive characteristics of ciclosporin A may correspond to its effects on prolactin and its receptor. Recently, the rat liver prolactin receptor has been cloned and sequenced [73]. The result of this sequence analysis has been important in demonstrating related cDNA in lymphocytes. Using the polymerase chain reaction and prolactin receptor-specific oligo primers, the existence of prolactin receptor cDNAs was demonstrated when mRNA was isolated from the Nb2 T cell line, thymocytes and splenocytes and used as a template [66].

Somatostatin Receptors

Somatostatin (SOM) affects many of the functional responses of lymphocytes: it (1) inhibits the proliferation of MOLT-4 lymphoblast and human T lymphocytes [74]; (2) blocks IL-2 receptor expression by human peripheral blood and lamina propria lymphocytes stimulated with phytohemagglutinin (PHA) [75], and (3) inhibits IgA production by murine splenic and Peyer's patch lymphocytes [76]. The suppressive qualities of SOM on immunoglobulin production is thought to be due to the activation of SOM receptors. These receptors are primarily found on phenotypically defined T suppressor cells (Lyt 2^+) of the spleen (94% positive for SOM receptor) and Peyer's patches (53% positive for SOM receptors) [77]. Initially, low affinity (K_d = 700 n*M*; B_{max} = 300–500 sites/cell) SOM receptors were identified on human peripheral blood mononuclear leukocytes [78]. However, subsequent studies showed that there are high affinity SOM receptors (K_d = 11 n*M*; $B_{max} = 7 \times 10^5$ sites/cell) on PHA-activated human lymphocytes but not on quiescent cells [79]. High affinity SOM receptors have also been found on the following cells: (1) Jurkat T cell line (K_d = 3 p*M*; 100 sites/cell); (2) U266 myeloma cells (K_d = 5 p*M*; 1,000 sites/cell [82], and (3) ($K_d = 2.1 \pm 0.3$ n*M*; $B_{max} = 11 \pm 1.4$ p*M*/7×10^5 cells) human intestinal lamina propria cells [81]. Activation of the SOM receptor by SOM inhibits cAMP levels in mouse spleen [86].

Substance P Receptors

Substance P (SP), a potent vasodilator, is thought to be involved in the pathogenesis of neurogenic inflammation [83] and rheumatoid arthritis [84]. Its effects on immunocompetence are well documented and most of the biological properties of SP probably act through the carboxy-terminal domain of the 11-amino acid neurotransmitter [85]. The SP receptor has been well characterized. SP receptors have been identified with high affinity characteristics from the following cells and cell lines: (1) human T lymphocytes (K_d = 400 p*M*; B_{max} = 7,000 sites/cell); (2) murine splenic T cells

(K_d = 620 p*M*; B_{max} = 195 sites/cell); (3) murine Peyer's patch T cells (K_d = 500 p*M*; B_{max} = 647 sites/cell); (4) murine splenic B lymphocytes (K_d = 640 p*M*; B_{max} = 190 sites/cell); (5) murine Peyer's patch B lymphocytes (K_d = 520 p*M*; B_{max} = 975 sites/cell), and (6) human IM-9 B lymphoma cells (K_d = 870 p*M*; B_{max} = 23,000 sites/cell) [86]. The results of ligand-specific chemical cross-linking studies using ^{125}I-SP show the SP receptor is a single polypeptide chain with a molecular weight of 58 kDa [87]. Immuno-affinity purification studies using an anti-SP receptor antibody have confirmed the molecular weight of the SP receptor [88].

Thyrotropin Receptor

Thyrotropin (TSH) is an anterior pituitary glycoprotein hormone which enhances in vitro T-dependent and T-independent antibody responses [89, 90]. TSH receptors have recently been identified on lymphoid cell lines and murine splenocytes [91]. Specifically, three cell lines including 702/3 pre-B cells, RPMI-1788 IgM-secreting B cells, and TIB190 IgG-secreting lymphoblasts have all been shown to posses high affinity (K_d = 1–3 n*M*; B_{max} = 25–50 fmol/mg) binding sites for TSH. The murine splenocyte TSH receptor is specific to B lymphocytes and its expression is directly related to the activation state of the cell. For example, murine T- or B-enriched splenic lymphocytes display no measurable TSH receptors [91]. However, LPS-stimulated, B-enriched cells specifically bind TSH with an IC_{50} of 1 n*M* while staphylococcal enterotoxin A (SEA) T-enriched lymphocytes do not specifically label TSH [91]. Since SEA elicits lymphocyte TSH production [92], it may be possible the lymphocyte-derived TSH is either (1) blocking the expression of the receptor at the translational or transcriptional level or (2) binding to the receptor precluding its detection by exogenous radiolabeled TSH. Apparently, TSH effects are mediated through a cAMP-dependent second messenger pathway since TSH stimulated immunoglobulin production in TIB190 lymphoblasts parallels increases in intracellular cAMP levels [91].

Vasoactive Intestinal Peptide Receptor

The vasoactive intestinal peptide (VIP) receptor is one of the best characterized neuropeptide receptors on cells of the immune system. High affinity sites (K_d = 200–500 p*M*) have been identified on human monocytes

(9,600 sites/cell) [93], human T-enriched lymphocytes (1,700 sites/cell) [94], human T-depleted lymphocytes (3,000 sites/cell) [95], human B-enriched lymphocytes (600 sites/cell) [96], and human LGLs (2,400 sites/cell) [100]. Likewise, human lymphoid cell lines of T- and B-cell lineage including Jurkat T cells (K_d = 70 p*M*) [97] and Dakiki (K_d = 90 p*M*) and Nalm-6 (K_d = 130 p*M*) B cells [98] possess VIP receptors. Similarly, rodent lymphocytes from spleen, mesenteric lymph nodes, and Peyer's patches have all been shown to possess high affinity (K_d = 100–200 p*M*; B_{max} = 500–3,000 sites/cell) VIP receptors [95, 99]. VIP regulates immunocompetence in vitro including antibody production, natural killer activity, histamine production by mast cells, and lymphocyte proliferation [100]. Moreover, in vivo injection of VIP-treated T lymphocytes alters their migration to mesenteric lymph nodes and Peyer's patches [100], indicating VIP is involved in the routing of lymphocyte migration.

Studies on MOLT-4b T-lymphoblastoid, Nalm6 pre-B, and Dakiki plasma cell lines show the receptor binding site has an apparent molecular weight of 47 kDa [98, 101]. Receptor activation stimulates the guanine nucleotide binding protein complex and the catalytic subunit of adenylate cyclase, resulting in the generation of cAMP and the activation of a cAMP-dependent protein kinase [100]. This was demonstrated in studies on MOLT-4b cells in which VIP treatment resulted in the phosphorylation of a 38-kDa protein [102]. VIP_{1-28} and variants thereof are apparently synthesized by immune cells predominately by granulocytes [103]. Cleavage products of VIP_{1-28} including VIP_{23-28}, VIP_{15-28}, and most likely VIP_{4-28} may potentially be endogenous antagonists of VIP, since they bind to the receptor but do not activate the cAMP pathway [80]. Evidence showing VIP-ergic nerve fibers innervating Peyer's patches indicate VIP may achieve levels locally which can significantly affect immune homeostasis [100]. In a reciprocal manner, VIP production by immune cells may act in turn as a neurophysiological signal to the nervous system resulting for example in alterations in electrolyte transport systems [80].

Other Neuropeptide Receptors

Other receptors have been identified on immune cells; however, their characterization is much less complete. For example, α-MSH affects immune cells by acting as an antipyretic anti-inflammatory molecule [104] and has recently been shown to antagonize IL-1, tumor necrosis factor, and C5a-mediated neutrophil chemotaxis [105]. These effects may be receptor-mediated since there is specific competitive uptake of $[^{125}I]$NDP-MSH by spleen cells but not thymocytes [106]. Another example is calcitonin gene-related peptide I (CGRP-I) and (CGRP-II) which have been shown to stimulate adenylate cyclase in the murine P388d_1 macrophage cell line [107]. $[^{125}I]$CGRP specifically labels P388d_1 membranes with an $IC_{50} \sim 7$ n*M* whereas VIP and calcitonin do not displace radiolabeled CGRP, suggesting that these cells possess a specific receptor for calcitonin gene-related peptide.

Hypothalamic Releasing Hormone Receptors

Cells of the immune system also respond to hypothalamic releasing hormone peptides in a manner remarkably similar to pituitary target cells. For example, CRH induces ACTH production by leukocytes [108]. In addition to stimulating hormone production, CRH also influences parameters associated with immunocompetence. For example, CRH augments IL-2 expression on T lymphocytes [109], stimulates B cell proliferation [110], and enhances natural killer activity (in the presence of AVP) by murine splenocytes [111]. CRH-mediated effects take place through specific cell surface receptors on immune cells. Some of the CRH-mediated events are due to the production of endorphins by immune cells. Specifically, CRH-mediated enhancement of natural killer activity is elicited through the production of opioids (presumably endorphin) which can then interact with the δ-class opioid receptor [111]. Cell surface CRH receptors are found on macrophage-enriched populations [112] and B lymphocytes [110]. The high affinity ($K_d = 260$ p*M*; $B_{max} = 8.74$ fmol/mg protein) receptor is coupled to the adenylate cyclase system as well [113].

Growth hormone-releasing hormone (GHRH) receptors have also been identified on cells of the immune system. The receptor binding site is saturable and found on both thymocytes ($K_d = 3.5$ n*M*; $B_{max} = 54$ fmol/10^6 cells) and splenic lymphocytes ($K_d = 2.5$ n*M*; $B_{max} = 35$ fmol/10^6 cells) [114]. The results of chemical cross-linking studies using $[^{125}I]$GHRH show that thymocytes possess two polypeptide binding sites with molecular weights of 42 and 27 kDa [114]. Following GHRH binding to its receptor, there is a rapid increase in intracellular Ca^{2+} which is associated with the stimulation of lymphocyte proliferation.

Luteinizing hormone-releasing hormone receptors have also been identified on thymocytes ($K_d = {=}84$ n*M*; $B_{max} = 14$ fmol/mg) [115]. These receptors are known to be involved in the development of the thymus and the maturation of the systemic immune system. Specifically, administration of the LHRH antagonist *p*-Glu-*D*-$Phe^{2,6}$, Pro^{3}-LHRH into neonatal female rats results in a dramatic loss in the ability of these animals to mount an immune response to bovine serum albumin [116]. Moreover, there is also a reduction in the number of thymocytes that respond to mitogen and a diminution in T helper cell numbers [117]. Immune cells respond to LHRH by producing LH [117] which occurs in both a time- and dose-dependent manner [118]. Lastly, structural studies on this receptor using immunoaffinity chromatography techniques have shown that the binding site is a single polypeptide with a molecular weight of 51 kDa [119]. Leukocytes also respond to thyrotropin-releasing hormone (TRH) treatment of producing TSH which is detectable at both the mRNA and protein levels [90]. This interaction is also thought to occur through a cell surface TRH receptor on lymphocytes.

In Summary

The functional role of neuropeptide receptors in the immune system seems to be sensory and regulatory. Signaling molecules (i.e. neuropeptides/neurotransmitters) secreted by immune or neuroendocrine cells can interact with specificity to immunocyte membrane bound neuroendocrine peptide receptors resulting in changes in immune homeostasis. At the onset of this review, three criteria were established in order to formally define the existence of a receptor including the pharmacologic, biochemical, and biological profiles. Applying these criteria, at least ten neuroendocrine peptide/peptide neurotransmitter receptors have met the requirements (table 1). Perhaps more striking, many of these receptors have characteristics which are nearly identical to those receptors found on neuroendocrine tissue. This observation implies irregularities in the physiology of a receptor system in one tissue compartment (neuroendocrine or immune) may be mirrored by the receptors found in the other compartment (neuroendocrine or immune) as well. This rather hypothetical concept is, in fact, supported by data studying ACTH [120], CRH [121], and TRH [122] receptors on immune cells relative to the expression or function of these receptors in neuroendocrine tissue taken from patients with neuroendocrine disorders. It is likely that future work will use this relationship to further study the dynamic interaction between the immune and neuroendocrine systems in defining neurologic, neuroendocrine, and autoimmune disorders.

Table 1. Neuroendocrine peptide receptors on cells of the immune system[a]

Receptor	Immunocyte source	K_d; B_{max}	M_r^b, kDa	Ref.
ACTH	Rat splenic T cells	K_{d_1} = 87 p*M*; 1,100 sites/cell	83	15
		K_{d_2} = 4.0 n*M*; 30,000 sites/cell		
	Rat splenic B cells	K_{d_1} = 90 p*M*; 3,600 sites/cell		
		K_{d_2} = 4.3 n*M*; 38,000 sites/cell		
β-Endorphin	EL-4 murine thymoma	65 n*M*	72	21–24
	Murine splenocytes	4.1 n*M*; 0.1–0.3 fmol/10^6 cells	57	
	U937 human promonocytic cell line	12 n*M*; 40 fmol/10^6 cells	66, 59, 44	
Nerve growth factor	Human B lymphocytes	1–10 n*M*; 4 fmol/10^8 cells	85	45, 46
	Human T lymphocytes	1–10 n*M*; 3 fmol/10^8 cells		
	Human monocytes	1–10 n*M*; 1 fmol/10^8 cells		
	Rat splenocytes	2.5 n*M*; 13,000 sites/cell	125, 190	
μ-Opioid	Rat and murine splenocytes	960 n*M*; 54 n*M*/2 × 10^7 cells	58–70	54, 56
κ-Opioid	EL-4 murine thymoma	60 n*M*; 2.7 pmol/10^6 cells		51
	$P388d_1$ murine macrophage cell line	17 n*M*; 54 fmol/10^6 cells	38–42	52, 57
Substance P	Human T cells	400 p*M*; 7,000 sites/cell		86–88
	Murine splenic T cells	620 p*M*; 195 sites/cell		
	Murine Peyer's patch T cells	500 p*M*; 647 sites/cell		
	Murine splenic B cells	640 p*M*; 190 sites/cell		
	Murine Peyer's patch B cells	920 p*M*; 975 sites/cell		
	Human IM-9 B lymphoma cells	870 p*M*; 23,000 sites/cell	58	

Table 1. (continued)

Receptor	Immunocyte source	K_d; B_{max}	M_r^b, kDa	Ref.
VIP	Human monocytes	250 p*M*; 9,600 sites/cell		98, 100, 101
	Human T-enriched cells	470 p*M*; 1,700 sites/cell		
	Human B-enriched cells	430 p*M*; 1,400 sites/cell		
	Human large granular lymphocytes	390 p*M*; 2,400 sites/cell		
	Human MOLT-4b T lymphoblastoid cell	8 n*M*;	50	
	Human pre-B Nalm-6 cells	13 n*M*; 2.6 n*M*/150 μg	47	
	Human Dakiki plasma cells	9 n*M*; 2.6 n*M*/150 μg	47	
	Rat splenocytes	100 p*M*; 1,600 sites/cell		
	Murine splenocytes	220 p*M*; 900 sites/cell		
	Murine Peyer's patch lymph.	240 p*M*; 500 sites/cell		
GHRH	Rat thymocytes	3.5 n*M*; 54 fmol/10^6 cells	27, 42	114
	Rat splenocytes	2.5 n*M*; 35 fmol/10^6 cells		
LHRH	Rat thymocytes	84 n*M*; 14 fmol/mg		115, 119
	Murine splenocytes		51	

[a] In order to meet the criteria establishing neuroendocrine peptide receptor expression on cells of the immune system, receptors were defined by pharmacological, biochemical (structural), and biological (functional) techniques.
[b] Mr = Apparent molecular weight (in kilodaltons) as determined by polyacrylamide gel electrophoresis.

References

1 Grantham JJ, Burg MB: Effect of vasopressin and cyclic AMP on permeability of isolated collecting tubules. Am J Physiol 1966;211:255–259.
2 Hems DA, Whitton PD: Stimulation by vasopressin of glycogen breakdown and gluconeogenesis in the perfused rat liver. Biochem J 1973;136:705–709.
3 Fishman JB, Dickey BF, Fine RE: Purification and characterization of the rat liver vasopressin (V_1) receptor. J Biol Chem 1987;262:14049–14055.
4 Fahrenholz F, Kajro E, Muller M, Boer R, Lohr R, Grzonka Z: Iodinated photoreactive vasopressin antagonist: Labelling of hepatic vasopressin receptor subunits. Eur J Biochem 1986;161:321–328.
5 Johnson HM, Torres BA: Regulation of lymphokine production by arginine vasopressin and oxytocin: Modulation of lymphocyte function by neurohypophyseal hormones. J Immunol 1985;135:773s–775s.
6 Johnson HM, Torres BA: Immunoregulatory properties of neuroendocrine peptide hormones; in Blalock JE, Bost KL (eds): Neuroimmunoendocrinology. Prog Allergy. Basel, Karger, 1988, vol 43, pp 37–67.
7 Elands J, Van Woundenberg A, Resink A, de Kloet ER: Vasopressin receptor capacity of human blood peripheral mononuclear cells is sex dependent. Brain Behav Immun 1990;4:30–38.
8 Torres BA, Johnson HM: Arginine vasopressin-binding peptides derived from the bovine and rat genomes differ in their abilities to block arginine vasopressin modulation of murine immune function. J Neuroimmunol 1990;27:191–199.
9 Vale W, Vaughan J, Smith M, Tamamoto G, Rivier J, Rivier C: Effects of synthetic ovine corticotropin-relasing factor, glucocorticoids, catecholamines, neurophypophyseal peptides, and other substances on cultured corticotropic cells. Endocrinology 1983;113:1121–1131.
10 Johnson HM, Smith EM, Torres BA, Blalock JE: Neuroendocrine hormone regulation of in vitro antibody production. Proc Natl Acad Sci USA 1982;79:4171–4174.
11 Heijnen CJ, Zijlstra J, Kavelaars A, Crosset G, Ballieux R: Modulation of the immune response by POMC-derived peptides. I. Influence on proliferation of human lymphocytes. Brain Behav Immun 1987;1:284–291.
12 Alvarez-Mon M, Kehrl JH, Fauci AS: A potential role for adrenocorticotropin in regulating human B lymphocyte functions. J Immunol 1985;135:3823–3826.
13 Bost KL, Smith EM, Wear LB, Blalock JE: Presence of ACTH and its receptor on a B lymphocytic cell line. A possible autocrine function for a neuroendocrine hormone. J Biol Regul Homeostat Agents 1987;1:23–27.
14 Bost KL, Clarke BL, Xu J, Kiyono H, McGhee JR, Pascual D: Modulation of IgM secretion and H chain mRNA expression in CH12.LX.C4.5F5 B cells by adrenocorticotropic hormone. J Immunol 1990;145:4326–4331.
15 Clarke BL, Bost KL: Differential expression of functional adrenocorticotropic hormone receptors by subpopulations of lymphocytes. J Immunol 1989;143:464–469.
16 Johnson EW, Blalock JE, Smith EM: ACTH receptor-mediated induction of leukocyte cyclic AMP. Biochem Biophys Res Commun 1988;157:1205–1211.
17 Gallo-Payet N: Adrenocorticotropin (ACTH) receptors; in Kalimi MY, Hubbard JR (eds): Peptide Hormone Receptors. Berlin, de Gruyter, 1987, pp 287–333.

18 Hazum E, Chang K-J, Cuatrecasas P: Specific nonopiate receptors for β-endorphin. Science 1979;205:1033–1035.
19 Carr DJJ, Blalock JE: Receptors for neuroendocrine peptides on cells of the immune system; in Blalock JE, Martin J, MacLeod R, Scapagnini U (eds): Advances in Neuroendocrinimmunology. New York, Thieme Medical Publishers, 1990, pp 41–72.
20 Gilmore W, Weiner LP: The opioid specificity of beta-endorphin enhancement of murine lymphocyte proliferation. Immunopharmacology 1989;17:19–30.
21 Schweigerer L, Schmidt W, Teschemacher H, Wilhelm S: β-Endorphin: Interaction with specific nonopioid binding sites on EL-4 thymoma cells. Neuropeptides 1985;6:445–452.
22 Schweigerer L, Schmidt W, Teschemacher H, Gramich C: β-Endorphin: Surface binding and internalization in thymoma cells. Proc Natl Acad Sci USA 1985;82: 5751–5754.
23 Shahahi NA, Linner KM, Sharp BM: Murine splenocytes express a naloxone-insensitive binding site for β-endorphin. Endocrinology 1990;126:1442–1448.
24 Shahahi N, Peterson PK, Sharp B: β-Endorphin binding to naloxone-insensitive sites on a human mononuclear cell line (U937): Effects of cations and guanosine triphosphate. Endocrinology 1990;126:3006–3015.
25 Scheideler MA, Zukin RS: Reconstitution of solubilized delta-opiate receptor binding sites in lipid vesicles. J Biol Chem. 1990;265:15176–15182..
26 Borboni P, di Cola G, Sesti G, Marini MA, Del Porto P, Saveria M, Montani G, Lauro R, De Pirro R: β-Endorphin receptors on cultured and freshly isolated lymphocytes from normal subjects. Biochem Biophys Res Commun 1989;163:642–648.
27 Hemmick LM, Bidlack JM: β-Endorphin modulation of mitogen-stimulated calcium uptake by rat thymocytes. Life Sci 1987;41:1971–1978.
28 Simonds WF: The molecular basis of opioid receptor function. Endocr Rev 1988;9: 200–212.
29 Carr DJJ, Bubien JK, Woods WT, Blalock JE: Opioid receptors on murine splenocytes. Possible coupling to K^+ channels. Ann NY Acad Sci 1988;540:694–697.
30 Yeung CWT: Photoaffinity labeling of opioid receptor of rat brain membranes with ^{125}I(*D*-Ala^2, *p*-N_3-Phe^4-Met^5)enkephalin. Arch Biochem Biophys 1987;254: 81–91.
31 Shrewsbury MM, Reinhardt WO: Effect of pituitary growth hormone on lymphatic tissues, thoracic duct lymph flow, lymph protein, and lymphocyte output in the rat. Endocrinology 1959;65:858–861.
32 Kelley KW: The role of growth hormone in modulation of the immune response. Ann NY Acad Sci 1990;594:95–103.
33 Arrenbrecht S: Specific binding of growth hormone to thymocytes. Nature (Lond) 1974;252:255–257.
34 Lesniak MA, Gorden P, Roth J, Gavin JR III: Binding of ^{125}I-human growth hormone to specific receptors in human cultured lymphocytes. J Biol Chem 1974;249: 1661–1667.
35 Kiess W, Butenandt O: Specific growth hormone receptors on human peripheral mononuclear cells: Reexpression, identification, and characterization. J Clin Endocrinol Metabl 1985;60:740–746.
36 Jafari P, Khansari DN: Detection of somatotropin receptors on human monocytes. Immunol Lett 1990;24:199–202.

37 Weigent DA, Baxter JB, Wear WE, Smith LR, Bost KL, Blalock JE: Production of immunoreactive growth hormone by mononuclear leukocytes. FASEB J 1988;2: 2812–2818.
38 Weigent DA, Blalock JE: The production of growth hormone by subpopulations of rat mononuclear leukocytes. Cell Immunol 1991;135:55–65.
39 Carr DJJ, Weigent DA, Blalock JE: Hormones common to the neuroendocrine and immune systems. Drug Design Del 1989;4:187–195.
40 Nagy E, Berczi I, Friesen HG: Regulation of immunity in rats by lactogenic and growth hormones. Acta Endocrinol 1983;102:351–357.
41 Leung DW, Spencer SA, Cachianes G, Hammonds RG, Collins C, Henzel WJ, Barnard R, Waters MJ, Wood WI: Growth hormone receptor and serum binding protein: Purification, cloning, and expression. Nature 1987;330:537–543.
42 Levi-Montalcini R, Aloe L, Alleva E: A role for nerve growth factor in nervous, endocrine, and immune system. Prog NeuroEndocrinImmunology 1990;3:1–11.
43 Stead RH, Tomioka M, Pezzati P, Marshall J, Coituru K, Perdue M, Stanisz A, Bienenstock J: Interaction of the mucosal immune and peripheral immune systems; in Ader R, Felten DL, Cohen N (eds): Psychoneuroimmunology. Orlando, Academic Press, 1991, pp 177–207.
44 Thorpe LW, Jerrells TR, Perez-Polo JR: Mechanisms of lymphocyte activation by nerve growth factor. Ann NY Acad Sci 1990;594:78–84.
45 Otten U, Ehrhard P, Peck R: Nerve growth factor induces growth and differentiation of human B lymphocytes. Proc Natl Acad Sci USA 1989;86:10059–10063.
46 Thorpe LW, Stach RW, Hashim GA, Marchetti D, Perez-Polo JR: Receptor for nerve growth factor on rat spleen mononuclear cells. J Neurosci Res 1987;17:128–134.
47 Wybran J, Appelboom T, Famaey J-P, Govaerts A: Suggestive evidence for receptors for morphine and methionine-enkephalin on normal human blood T lymphocytes. J Immunol 1979;123:1068–1070.
48 McDonough RJ, Madden JJ, Falek A, Shafer DA, Pline M, Gordon D, Bokos P, Kuehnle JC, Mendelson J: Alteration of T and null lymphocyte frequences in the peripheral blood of human opiate addicts: In vivo evidence for opiate receptor sites on T lymphocytes. J Immunol 1980;125:2539–2543.
49 Lopker A, Abood LG, Hoss W, Lionetti FJ: Stereoselective muscarinic acetylcholine and opiate receptors in human phagocytic leukocytes. Biochem Pharmacol 1980;29: 1361–1365.
50 Ausiello CM, Roda LG: Leu-enkephalin binding to cultured human T lymphocytes. Cell Biol Int Rep 1984;8:97–106.
51 Fiorica E, Spector S: Opioid binding site in EL-4 thymoma cell line. Life Sci 1988;42:199–206.
52 Carr DJJ, DeCosta BR, Kim C-H, Jacobson AE, Guarcello V, Rice KC, Blalock JE: Opioid receptors on cells of the immune system: Evidence for δ- and κ-classes. J Endocrinol 1989;122:161–168.
53 Carr DJJ, Blalock JE: Neuroendocrine characteristics of the immune system. EOS J Immunol Immunopharmacol 1989;IX:195–199.
54 Ovadia H, Nitsan P, Abramsky O: Characterization of opiate binding sites on membranes of rat lymphocytes. J Neuroimmunol 1989;21:93–102.
55 Carr DJJ, Kim C-H, DeCosta B, Jacobson AE, Rice KC, Blalock JE: Evidence for a δ-class opioid receptor on cells of the immune system. Cell Immunol 1988;116:44–51.

56 Radulescu RT, DeCosta BR, Jacobson AE, Rice KC, Blalock JE, Carr DJJ: Biochemical and functional characterization of a mu-opioid receptor binding site on cells of the immune system. Prog NeuroEndocrinImmunology 1991;4:166–179.

57 Carr DJJ, DeCosta BR, Jacobson AE, Rice KC, Blalock JE: Enantioselective kappa opioid binding sites on the macrophage cell line, P388d$_1$. Life Sci 1991;49:45–51.

58 Fulop T Jr, Kekessy D, Foris G: Impaired coupling of naloxone sensitive opiate receptors to adenylate cyclase in PMNLs of aged male subjects. Int J Immunopharmacol 1987;9:651–658.

59 Carr DJJ, Bost KL, Blalock JE: The production of antibodies which recognize opiate receptors on murine leukocytes. Life Sci 1988;42:2615–2624.

60 Hough CJ, Halperin JI, Mazorow DL, Yeandle SL, Millar DB: β-Endorphin modulates T-cell intracellular calcium flux and c-myc expression via a potassium channel. J Neuroimmunol 1990;27:163–171.

61 Shavit Y, DePaulis A, Martin FC, Terman GW, Pechnick RN, Zane CJ, Gale RP, Liebeskind JC: Involvement of brain opiate receptors in the immune-suppressive effect of morphine. Proc Natl Acad Sci USA 1986;83:7114–7117.

62 Weber RJ, Pert A: The periaqueductal graph matter mediates opiate-induced immunosuppression. Science 1989;245:188–190.

63 Bryant HV, Roudebush RE: Suppressive effects and morphine pellet implants on in vivo parameters of immune function. J Pharmacol Exp Ther 1990;255:410–414.

64 Stein C, Hassan AHS, Przewlocki R, Gramsch C, Peter K, Herz A: Opioids from immunocytes interact with receptors on sensory nerves to inhibit nociception in inflammation. Proc Natl Acad Sci USA 1990;87:5935–5939.

65 Bernton EW, Meltzer MS, Holaday JW: Suppression of macrophage activation and T-lymphocyte function in hypoprolactinemic mice. Science 1988;239:401–404.

66 Yu-Lee L-Y, Stevens AM, Hrachovy JA, Schwarz LA: Prolactin-mediated regulation of gene transcription in lymphocytes. Ann NY Acad Sci 1990;594:146–155.

67 Gerli R, Rambotti P, Nicoletti I, Orlandi S, Migliorati G, Riccardi C: Reduced number of natural killer cells in patients with pathological hyperprolactinemia. Clin Exp Immunol 1986;64:399–406.

68 Russell DH, Matrisian L, Kibler R, Larson DF, Poulos B, Magun BE: Prolactin receptors on human lymphocytes and their modulation by cyclosporin. Biochem Biophys Res Commun 1984;121:899–906.

69 Matera L, Muccioli G, Cesano A, Bellussi G, Genazani E: Prolactin receptors on large granular lymphocytes: Dual regulation by cyclosporin A. Brain Behav Immun 1988;2:1–10.

70 Russell DH, Kibler R, Matrisian L, Larson DF, Poulos B, Magun BE: Prolactin receptors on human T and B lymphocytes: Antagonism of prolactin binding by cyclosporine. J Immunol 1985;134:3027–3030.

71 Hartmann D, Holaday J, Bernton E: Inhibition of lymphocyte proliferation to PRL. FASEB J 1989;3:2194–2199.

72 Mukherjee P, Mastro AM, Hymer WC: Prolactin induction of interleukin-2 receptors on rat splenic lymphocytes. Endocrinology 1990;126:88–94.

73 Boutin J, Jolicoeur C, Okamura H, Gagnon J, Edery M, Shirota M, Banuille D, Dusanter-Fourt I, Dijiane J, Kelly P: Cloning and expression of the rat prolactin receptor, a member of the growth hormone/PRL receptor gene family. Cell 1988;53: 69–77.

74 Payan DG, Heis CA, Goetzl EJ: Inhibition by somatostatin of the proliferation of T-lymphocytes and MOLT-4 lymphoblasts. Cell Immunol 1984;84:433–438.

75 Pallone F, Fais S, Annibale B, Boirivant M, Morace S, Delle Fave G: Modulatory effects of somatostatin and vasoactive intestinal peptide on human intestinal lymphocytes. Ann NY Acad Sci 1990;594:408–410.
76 Stanisz AM, Befus D, Bienenstock J: Differential effects of vasoactive intestinal peptide, substance P, and somatostatin on immunoglobulin synthesis and proliferation by lymphocytes from Peyer's patches, mesenteric lymph nodes, and spleen. J Immunol 1986;136:152–156.
77 Scicchitano R, Dazin P, Bienenstock J, Payan DG, Stanisz AM: Distribution of somatostatin receptors on murine spleen and Peyer's patch T and B lymphocytes. Brain Behav Immun 1987;1:173–184.
78 Bhatena SJ, Lovie J, Scheitter GP, Redman RS, Wahl L, Recant L: Identification of human mononuclear leukocytes bearing receptors for somatostatin and glucagon. Diabetes 1981;30:127–134.
79 Hiruma K, Nakamura KH, Sumida T, Maeda T, Tomioka H, Yoshida S, Fujita T: Somatostatin receptors on human lymphocytes and leukaemia cells. Immunology 1990;71:480–485.
80 Goetzl EJ, Turck DW, Sreedharan SP: Production and recognition of neuropeptides by cells of the immune system; in Ader R, Felten DL, Cohen N (eds): Psychoneuroimmunology. Orlando, Academic Press, 1991, pp 263–282.
81 Fais S, Annibale B, Boirivant M, Santoro A, Pallone F, Delle Fave G: Effects of somatostatin on human intestinal lamina propria lymphocytes. Modulation of lymphocyte activation. J Neuroimmunol 1991;31:211–219.
82 Pawlikowski M, Stepien M, Kunter-Radek J, Shally AV: Effect of somatostatin on the proliferation of mouse spleen lymphocytes in vitro. Biochem Biophys Res Commun 1985;129:52–55.
83 Foreman JC: Peptides and neurogenic inflammation. Br Med Bull 1987;43:386–400.
84 Kimball ES: Substance P, cytokines, and arthritis. Ann NY Acad Sci 1990;594:293–308.
85 McGillis JP, Organist ML, Scriven KH, Payan DG: Purification of the 33,000-dalton ligand binding-protein constituent of the lympoblast substance P receptor. J Neurosci Res 1987;18:190–194.
86 McGillis JP, Mitsuhashi M, Payan DG: Immunomodulation by tachykinin neuropeptides. Ann NY Acad Sci 1990;594:85–94.
87 Payan DG, McGillis JP, Organist ML: Binding characteristics and affinity labeling of protein constituents of the human IM-9 lymphoblast receptor for substance P. J Biol Chem 1986;261:14321–14329.
88 Pascual DW, Blalock JE, Bost KL: Antipeptide antibodies that recognize a lymphocyte substance P receptor. J Immunol 1989;143:3697–3702.
89 Blalock JE, Johnson HM, Smith EM, Torres BA: Enhancement of the in vitro antibody response by thyrotropin. Biochim Biophys Res Commun 1984;125:30–34.
90 Kruger TE, Smith LR, Harbour DV, Blalock JE: Thyrotropin. An endogenous regulator of the in vitro immune response. J Immunol 1989;142:744–747.
91 Harbour DV, Leon S, Keating C, Hughes TK: Thyrotropin modulates B-cell function through specific bioactive receptors. Prog NeuroEndocrinImmunology 1990;3:266–276.
92 Smith EM, Phan M, Coppenhaver D, Kruger TE, Blalock JE: Human lymphocyte production of immunoreactive thyrotropin. Proc Natl Acad Sci USA 1983;80:6010–6013.

93 Wiik P, Opstad P, Boyum A: Binding of vasoactive intestinal polypeptide by human blood monocytes: Demonstration of specific binding sites. Regul Pept 1985;12:145–153.
94 Danek A, O'Dorisio M, O'Dorisio T, George J: Specific binding sites for vasoactive intestinal polypeptide on nonadherent peripheral blood lymphocytes. J Immunol 1983;131:1173–1177.
95 Cavo J, Molinero P, Jiminez J, Goberna R, Guerreo J: Interaction of vasoactive intestinal peptide with rat lymphoid cells. Peptides 1986;7:177–181.
96 Ottaway CA: The effect of ligand internalization on cellular binding studies of peptide ligands. Ann NY Acad Sci 1990;594:45–59.
97 Finch R, Sreedharan S, Goetzl E: High affinity receptors for vasoactive intestinal peptide on human myeloma cells. J Immunol 1989;142:1977–1981.
98 O'Dorisio MS, Shannon BT, Fleshman DJ, Campolito LB: Identification of high affinity receptors for vasoactive intestinal peptide on human lymphocytes of B cell lineage. J Immunol 1989;142:3533–3536.
99 Ottaway C, Greenberg G: Interaction of vasoactive intestinal peptide with mouse lymphocytes: Specific binding and modulation of mitogen responses. J Immunol 1984;132:417–423.
100 Ottaway C: Vasoactive intestinal peptide and immune function; in Ader R, Felten DL, Cohen N (eds): Psychoneuroimmunology. Orlando, Academic Press, 1991, pp 225–262.
101 Woods CL, O'Dorisio MS: Covalent cross-linking of vasoactive intestinal polypeptide to its receptors on intact human lymphoblasts. J Biol Chem 1985;260:1243–1247.
102 O'Dorisio MS, Campolito LB: Comparison of vasoactive intestinal peptide-mediated protein phosphorylation in human lymphoblasts and colonic epithelial cells. Mol Immunol 1989;26:583–590.
103 Wanger GD, O'Dorisio MS, Goetzl EJ: Vasoactive intestinal peptide: Messenger in a neuroimmune axis. Ann NY Acad Sci 1990;594:104–119.
104 Lipton JM: Neuropeptide α-melanocyte-stimulating hormone in control of fever, the acute phase response, and inflammation; in Goetzl EJ, Spector NJ (eds): Neuroimmune Networks: Physiology and Diseases. New York, Liss, 1989, pp 243–250.
105 Mason MJ, Van Epps DE: Modulation of interleukin-1, tumor necrosis factor and C5a-mediated neutrophil migration by alpha-melanocyte stimulating hormone (MSH). J Immunol 1989;142:1646–1651.
106 Tatro JB, Reichin S: Specific receptors for α-melanocyte-stimulating hormone are widely distributed in tissues of rodents. Endocrinology 1987;121:1900–1907.
107 Abello J, Kaiserlian-Nicolas D, Cuber JC, Revillard JP, Chayvialle JA: Identification of high affinity calcitonin gene-related peptide receptors on a murine macrophage-like cell line. Ann NY Acad Sci 1990;594:364–366.
108 Smith EM, Morrill AC, Meyer WJ III, Blalock JE: Corticotropin releasing factor induction of leukocyte-derived immunoreactive ACTH and endorphins. Nature 1986;322:881–882.
109 Singh VK: Stimulatory effect of corticotropin-releasing neurohormone on human lymphocyte proliferation and interleukin-2 receptor expression. J Neuroimmunol 1989;23:257–262.
110 McGillis JP, Park A, Rubin-Fletter P, Turck C, Dallman MF, Payan DG: Stimulation of rat B-lymphocyte proliferation by corticotropin-releasing factor. J Neurosci Res 1989;23:346–352.

111 Carr DJJ, DeCosta BR, Jacobson AE, Rice KC, Blalock JE: Corticotropin-releasing hormone augments natural killer cell activity through a naloxone-sensitive pathway. J Neuroimmunol 1990;28:53–61.
112 Webster EL, Tracey DE, Jutila MA, Wolfe SA Jr, De Souza EB: Corticotropin-releasing factor receptors in mouse spleen: Identification of receptor bearing cells as resident macrophages. Endocrinology 1990;127:440–452.
113 Webster EL, Battaglia G, De Souza EB: Functional corticotropin-releasing factor (CRF) receptors in mouse spleen: Evidence from adenylate cyclase studies. Peptides 1989;10:395–402.
114 Guarcello V, Weigent DA, Blalock JE: Growth hormone releasing hormone receptors on thymocytes and splenocytes from rats. Cell Immunol 1991;136:291–302.
115 Marchetti B, Guarcello V, Morale MC, Bartoloni G, Farinella Z, Cordaro S, Scapagnini U: Luteinizing hormone-releasing hormone-binding sites in the rat thymus: Characteristics and biological function. Endocrinology 1989;125:1025–1036.
116 Morale MC, Batticane N, Bartoloni G, Guarcello V, Farinella Z, Galasso MG, Marchetti B: Blockade of central and peripheral luteinizing hormone-releasing hormone (LHRH) receptors in neonatal rats with a potent LHRH-antagonist inhibits the morphofunctional development of the thymus and maturation of the cell-mediated and humoral immune responses. Endocrinology 1991;128:1073–1085.
117 Ebaugh MJ, Smith EM: Human lymphocyte production of immunoreactive luteinizing hormone (abstract). Fed Proc 1987;46:7811.
118 Blalock JE, Costa O: Immune neuroendocrine interactions: Implications for reproductive physiology. Ann NY Acad Sci 1989;564:261–266.
119 Costa O, Mulchahey JJ, Blalock JE: Structure and function of luteinizing hormone-releasing hormone (LHRH) receptors on lymphocytes. Prog NeuroEndocrinImmunology 1990;3:55–60.
120 Smith EM, Brosnan P, Meyer WJ III, Blalock JE: An ACTH receptor on human mononuclear leukocytes. N Engl J Med 1987;317:1266–1269.
121 Singh VK, Fudenberg HH: Binding of [^{125}I]corticotropin releasing factor to blood immunocytes and its reduction in Alzheimer's disease. Immunol Lett 1988;18:5–8.
122 Harbour DV, Anderson A, Farrington J, Wassef A, Smith EM, Meyer WJ III: Decreased mononuclear leukocyte TSH responsiveness in patients with major depression. Biol Psychiatry 1988;23:797–806.

Dr. Daniel J.J. Carr, Department of Microbiology and Immunology,
LSU Medical Center, 1901 Perdido Street, New Orleans, LA 70112–1393 (USA)

Blalock JE (ed): Neuroimmunoendocrinology, 2nd rev ed.
Chem Immunol. Basel, Karger, 1992, vol 52, pp 106–153

Cytokines: Influence on Glial Cell Gene Expression and Function

Etty N. Benveniste

Departments of Cell Biology and Neurology, University of Alabama at Birmingham, UAB Station, Birmingham, Ala., USA

Introduction

The central nervous system (CNS) has traditionally been considered as an 'immunologically privileged site' for two major reasons: (1) the CNS lacks for the most part a lymphatic system that drains the tissues and captures potential antigens, and (2) the CNS is protected from the blood by the blood-brain barrier (BBB), which is impermeable to many soluble substances (immunoglobulins, cytokines, growth factors), and restricts the migration of lymphoid cells into the CNS. The BBB is a layered structure of endothelial cells of cerebral vessels, the basement membrane, and the perivascular glia limitans [1]. Additionally, cells of the CNS (neurons, macroglia, microglia) express very low levels of antigens encoded for by major histocompatibility complex (MHC) genes, whose products play a fundamental role in the induction and regulation of immune responses [2]. However, recent studies have demonstrated that the separation of the brain from the immune system is not absolute even under normal conditions, i.e. an intact BBB. T cells in very low numbers are found within normal brain tissue, and also in normal cerebrospinal fluid (CSF) [3]. Also, the presence of lymphatic-like capillaries in the brain provides a possible natural, untraumatized, entryway for lymphoid cells into the CNS [4]. Thus, there may exist a low level immune response in the CNS as a mechanism to eliminate potential antigens. Pathological events within the CNS often result in a breakdown of the BBB, which permits cells of the peripheral immune system to enter this site. During human diseases such as viral encephalitis [5], multiple sclerosis (MS) [6–8], and AIDS dementia complex (ADC) [9], and animal models of CNS disease such as experimental allergic encephalomyelitis (EAE) [10], inflammatory infiltrates composed of activated T cells, B cells, and macrophages are found in the brain.

There is a growing body of literature which suggests that communication exists between cells of the immune system and the CNS. There is increasing evidence that soluble factors from lymphoid/mononuclear cells are able to modulate the growth and function of cells found within the CNS; specifically, macroglia and microglia cells. Furthermore, glial cells can secrete immunoregulatory molecules that influence immune cells, as well as the glial cells themselves. Thus, the potential exists for bidirectional communication between lymphoid cells and glial cells, which is mediated via soluble factors.

In this chapter, I will initially describe the 'traditional' functions of various glial cells, and discuss their cell lineage, function, and antigenic determinants. I will then focus on various neurological disease states in which glial cells themselves may be involved in inflammation and immunological events occurring in the brain. The topics to be discussed include the ability of glial cells to both respond to, and synthesize, a variety of cytokines within the CNS. The capacity of glial cells to acquire MHC antigens and function as antigen-presenting cells within the CNS will also be covered. The implications of these functions, cytokine secretion and antigen presentation, by glial cells, will be discussed with respect to intracerebral immune responses, demyelination, and inflammation in neurological diseases that have an immunological component.

Glial Cells: Classification, Lineage and Function

The first account of neuroglia was cited by Dutrochet in 1824, who noticed the existence in the CNS of nonneuronal components made up of spindle-shaped cells which were morphologically distinct from neurons. These cells were considered as a form of connective tissue within the CNS, and were called 'neuroglia', which means nerve glue [11]. There are two broad subgroups of glial cells: the macroglia, which consists of astrocytes, oligodendrocytes and ependymal cells, and microglia. The cells to be described and discussed in this chapter are mammalian astrocytes, oligodendrocytes and microglia.

Astrocytes

Astrocyte Lineage

Astrocytes are the largest, most numerous of the glial cells, and in the mammalian brain, outnumber neurons 10:1. There are two morphological types of astrocytes; the type 1 or protoplasmic astrocyte, which is found predominantly in gray matter and in relation to capillaries, and the type 2 or fibrous astrocyte, which is localized primarily in the white matter [1]. It has been proposed that type 1 and type 2 astrocytes represent two separate cell lineages, and at least in vitro, that they develop from distinct precursor cells [for review, see 12, 13]. Other differences between the type 1 and 2 astrocytes include (1) morphology, (2) antigenic phenotype, (3) kinetics of development and appearance, and (4) response to growth factors. The scheme for type 1 astrocyte development is as follows: *Stem cells,* uncommitted multipotential cells, arise from neuroepithelial cells. From the stem cells arise *glioblasts,* which are cells committed to the macroglial lineage. *Proastroblasts* are derived from glioblasts, and are proliferative cells which are committed to the astrocyte lineage. The *astroblast* represents a more differentiated, proliferative cell committed to the astrocyte lineage. The type 1, or *protoplasmic astrocyte* is the terminally differentiated cell of this lineage (fig. 1a). These fully mature cells are detected at embryonic day 15 in the rat. The phenotype of the type 1 astrocyte changes with development. In differentiating astrocytes, the most characteristic feature of the astrocyte cytoplasm are intermediate filaments. The major component of these filaments is glial fibrillary acidic protein (GFAP), a 49,000-dalton protein unique to astrocytes [14]. Both the astroblast and type 1 astrocyte are GFAP-positive. A monoclonal antibody to an antigen consisting of certain gangliosides (A2B5) [15] detects a cell type termed the 'common oligodendrocyte-type 2 astrocyte progenitor cell' (0–2A), which will be discussed in more detail later, as well as neurons [16]. This antigen is not found on type 1 astrocytes, nor on its progenitor cells [17].

Type 2 fibrous astrocytes arise from A2B5-positive *0-2A bipotential progenitor cells,* which are detected at embryonic day 21 in the rat. The 0-2A progenitor cell can give rise to either the type 2 astrocyte or oligodendrocyte, depending on the environmental influence. In vitro, the 0-2A progenitor cell, in the presence of fetal calf serum, differentiates into a type 2 astrocyte, and in the absence of serum, into an oligodendrocyte [16]. It has been speculated that oligodendrocyte differentiation is the constitutive pathway of 0-2A progenitor cell development, whereas type 2 astrocyte differentiation is an induced pathway. Recent evidence suggests that ciliary neurotrophic factor can initiate type 2 astrocyte differentiation, but that other signals are required to drive the process to completion [18]. The phenotype of the mature type 2 astrocyte is as follows: A2B5-positive, thus retaining the surface antigen of its progenitor cell, and GFAP-positive (fig. 1b). In the rat

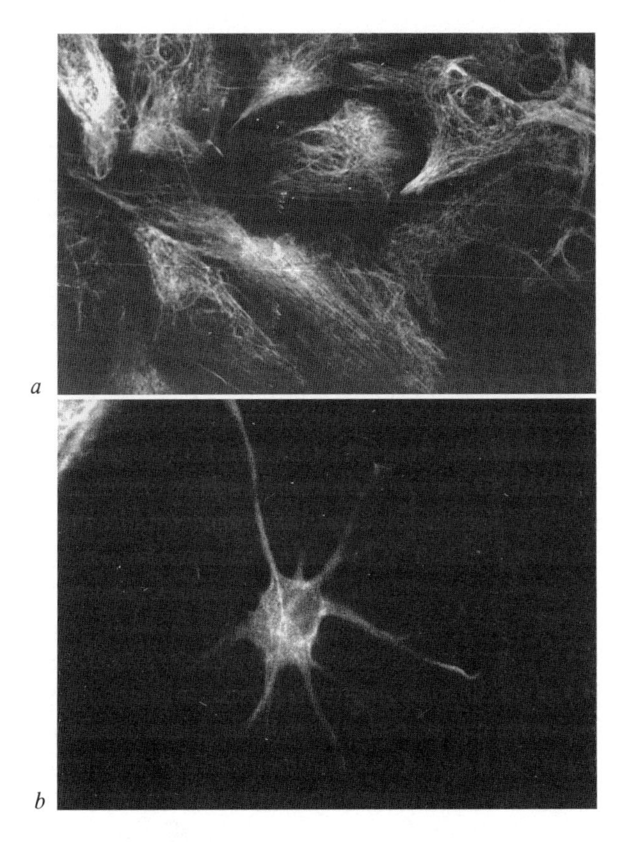

Fig. 1. Photomicrograph of rat protoplasmic *(a)* and fibrous *(b)* astrocytes. Primary rat astrocytes were isolated from mixed glial cultures on day 24, replated on coverslips at 5×10^3, and incubated for 4 days in serum-containing media. Cytoplasmic staining was performed on fixed cells with a monoclonal antibody to GFAP (1:4; Boehringer Mannheim) for 30 min at room temperature, followed by incubation with FITC-conjugated goat-anti-mouse IgG (1:20; Southern Biotechnology). *a* ×25. *b* ×50.

system, type 2 fibrous astrocytes are found predominantly in white matter by days 7–10 postnatally.

Astrocyte-Specific Antigens

Astrocytes are identified by a variety of antigens, the most predominant and specific being GFAP. GFAP is thought to contribute to maintaining the distinct morphology of the astrocyte. Glutamine synthetase (GS) is a glial enzyme that has been localized to astrocytes [19]. The function of this enzyme is to regulate the levels of the neurotransmitters glutamate and γ-aminobutyric acid (GABA), and also to participate in ammonia detoxification in the CNS [20]. Levels of GS can be increased by hydrocortisone administration both developmentally and in adult brain [21, 22]. S-100 protein is localized primarily in astrocytes [23, 24], and increases in concentration as the astrocyte matures [25]. S-100 is thought to be related to some of the Ca^{2+} binding proteins [26], and may function for cation transport.

Astrocyte Function

The astrocyte (type 1 protoplasmic) is the most versatile cell within the CNS, and has a variety of functions, several of which will be briefly described:

Mechanical support of neurons: The astrocyte provides a nonrigid supporting and insulating matrix for neurons. Because of their large numbers in the CNS, and the strength of the cytoplasmic projections (due to intermediate filaments), the astrocyte provides a supportive framework for nerve cells within the CNS.

Regulation of ions and metabolites: The ionic composition of the extracellular space around the neurons is critical, and the astrocyte is important in maintaining the proper environment. The presence of astrocytic end-feet on capillaries in the CNS have implicated the astrocyte in the regulation of water and ion balance in the CNS [27]. In particular, the flow of positively charged sodium and potassium ions are responsible for the generation of action potentials, so the levels of these ions must be finely regulated around the neurons. Some astrocytes have a high affinity for potassium [28], and function to remove excess potassium by a process called 'potassium spacial buffering'. Astrocytes also have several transport systems: one for the coupled exchange of sodium ions and hydrogen ions, and one for the exchange of chloride and bicarbonate ions, which allows the astrocyte to exchange these substances with neighboring cells, including neurons. The levels of ammonia are closely regulated in the brain, and the enzyme GS is involved in its metabolism.

Neurotransmitter transport system: Numerous studies have demonstrated the involvement of astrocytes in neurotransmitter metabolism. One of the best studied systems involves the metabolism of GABA. GABA is an inhibitor of neuronal transmission, and is metabolized by the enzyme GABA transaminase [29]. Astrocytes have a high-affinity uptake and metabolism of GABA, have high levels of GABA transaminase, and are thought to be physiologically important for removing GABA from synaptic clefts [30, 31].

As mentioned above, astrocytes are also involved in the metabolism of glutamate, via the enzyme GS. This enzyme is confined to astrocytes, and astrocytes in areas of the brain with high glutamergic activity show the heaviest concentration of GS staining by immunohistochemistry [32].

Guidance of migrating neurons during development: Another function performed by glia is assistance in ensuring the achievement of the correct connective relationships between neurons during development. Migrating neurons use radial glial cells to aid in their placement [33]. Radial glia are astrocyte-related cells which appear during brain development, and then become transformed, presumably into GFAP-positive type 1 astrocytes [34]. These cells are elongated, often bipolar, and have two or more principal cell processes extending long distances through neural tissue. In vivo, immature migrating neurons will cling closely to radial glia, and migrate along them to their final destination. It is thought that radial glia express adhesion molecules specific for developing neurons that allows them to arrive at their appropriate location in the brain [35].

Induction of the BBB: The capillary endothelial cells that make up the BBB are joined by tight junctions; as a result, the BBB is virtually impermeable to soluble substances. The CNS capillaries are almost completely surrounded by astrocytic end-feet, thus, the astrocyte can contribute to the structural integrity of the BBB. Recent evidence suggests that the formation of tight junctions between the endothelial cells is induced by astrocytes contacting the endothelium [36].

Astroglial response to injury: In the adult nervous system, astrocytes, unlike neurons, retain the ability to divide and multiply. When the CNS is injured, astrocytes respond by becoming reactive. The astrocytes proliferate, hypertrophy (enlargement of the cell body), and express more GFAP [37]. This reaction, known as fibrous gliosis or *astrogliosis,* eventually produces dense glial scars in the CNS. Depending on the circumstances, this gliotic scar can be viewed as either beneficial, such as walling off an abscess, or detrimental, such as in the disease of MS, where the formation of these scars is implicated in the clinical manifestations of this disease [38]. The significance of reactive astrocytes will be discussed later on in the context of inflammatory infiltration into the CNS.

Immunocompetent cell in the CNS: Recent studies have demonstrated that astrocytes may be involved in immunological events occurring in the brain. The astrocyte, upon stimulation, can be induced to express MHC antigens and secrete the cytokines IL-1, IL-6, TNF-α, and colony-stimulating factors (CSFs), all molecules that stimulate the growth and differentiation of macrophages and lymphoid cells. As such, the astrocyte has several important functional characteristics unique to traditional antigen-presenting cells (APC), demonstrating the astrocyte's potential to act as an immunocompetent cell in the CNS. These attributes will be discussed in detail in this chapter.

Oligodendrocytes

Oligodendrocyte Lineage

Oligodendrocytes, like astrocytes, arise from neuroepithelial cells. In vitro, oligodendrocytes arise from the 0-2A progenitor cell, which is detected at embryonic day 21 (in the rat) [16]. As mentioned previously, this progenitor cell can differentiate into either a type 2 astrocyte or oligodendrocyte, depending on environmental influences. In vitro, most oligodendrocytes are generated in the period of 7–17 days postnatally, with the peak occurring at day 14 [39].

Oligodendrocyte-Specific Antigens

Oligodendrocytes, the myelin-producing cells of the CNS, are identified for the most part by antisera directed against myelin-specific components. Myelin is composed of 70–85% lipid and 15–30% protein [40, 41]. The most commonly used marker for oligodendrocyte identification is galactocerebroside (GalC), the major glycolipid of myelin. This antigen is unique to oligodendrocytes, and localized on the cell surface (fig. 2). [42] A second myelin-specific glycolipid is sulfatide, which is the sulfated analog of GalC [43].

The major proteins of myelin include myelin basic protein (MBP), which makes up 30–35% of total myelin protein, and proteolipid protein (PLP), which accounts for 50% of total myelin protein [41]. MBP is associated with the cytoplasmic face of the myelin membrane, and is thought to mediate compaction between adjacent cytoplasmic membrane surfaces. PLP is an integral membrane protein thought to mediate interactions between opposing extracellular membrane surfaces. Myelin-associated glycoprotein (MAG) is closely associated with myelin, and makes up only 1% of total myelin proteins [41]. Recent data suggests that MAG may act as a cell-adhesion molecule [44]. The expression of all three myelin proteins, MBP, PLP and MAG, in tissue culture is indicative of oligodendrocyte maturation [45].

Oligodendrocytes also have several cell-specific enzymes that are thought to be involved in the myelination process. High activity of 2′,3′-

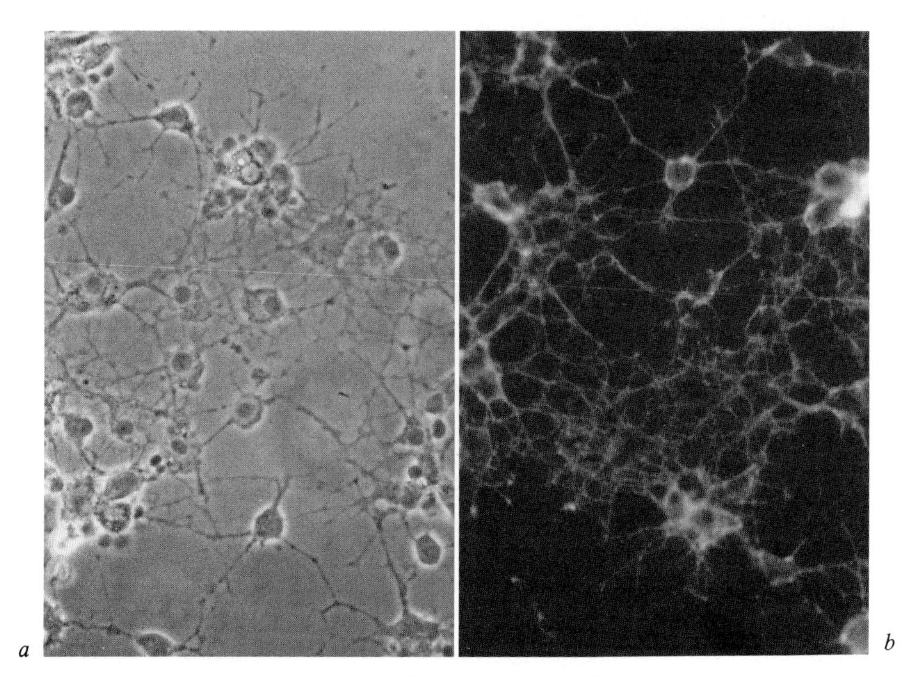

Fig. 2. Photomicrograph of rat oligodendrocytes. Oligodendrocytes were isolated from mixed glial cultures on day 10, replated on coverslips at 2×10^4, incubated for 4 days, and then stained for galactocerebroside (GalC). Surface staining for GalC was performed for 30 min in the cold with rabbit anti-GalC antibody (1:25), followed by FITC-conjugated goat-anti-rabbit Ig (1:20; Cappel) for 30 min in the cold. Cells were then fixed in cold acetone for 10 s. *a* Phase contrast microscopy. *b* Immunofluorescence staining for GalC.

cyclic nucleotide 3′-phosphohydrolase (CNPase) (an enzyme that hydrolyzes the 3′-phosphodiester bond of 2′,3′-cyclic nucleotides to produce 2′-monophosphate nucleotides) is found in the CNS in myelin or myelin-oligodendrocyte structures [46]. The exact biological function of this enzyme is unknown, but developmental changes in CNPase activity suggests an involvement with myelination [47]. Glycerol-3-phosphate dehydrogenase (GPDH) is an enzyme involved in glucose metabolism. In the CNS, its activity is thought to be involved with myelin lipid biosynthesis [48]. In vitro, levels of GPDH in oligodendrocytes can be elevated by hydrocortisone treatment [49].

Table 1. Characteristics of macroglial cells in the CNS

	Type 1 astrocyte	0-2A bipotential progenitor cell	Oligodendro-cyte	Type 2 astrocyte
Time of first appearance (rat)	embryonic day 15	embryonic day 21	postnatal day 7–17	postnatal day 7–10
Antigenic phenotype				
GFAP	+	–	–	+
GalC	–	–	+	–
A2B5	–	+	–	+
MBP	–	–	+	–

Oligodendrocyte Function

The function of the oligodendrocyte is myelin formation in the CNS. Myelin is wrapped around axons, acts as insulation for nerve fibers, and allows for efficient nerve impulse conduction [50]. Cytoplasmic projections extend from the oligodendrocyte cell body to wrap around nerve fibers in a spiral fashion. The oligodendrocyte is capable of producing many inter-modes of myelin simultaneously; in rat optic nerve, a single oligodendrocyte can myelinate up to 50 separate axons [51]. See table 1 for a summary of macroglial cell characteristics in the CNS.

Microglia

Microglia constitute approximately 10% of the total glial cell population. They are considered as the resident macrophages of the brain [for review, see 52]. Many different names have been given to microglia because of their variable morphologic appearance. The major subtypes of microglia include ramified, ameboid and perivascular microglia. *Ramified microglia* appear as a highly branched small cell, with branching occurring in all planes. The branching of microglial cell processes is often found around neurons, suggesting that there may be a functional significance to this physical association. *Perivascular microglia* are found in the perivascular space, and are thought to be more closely related to monocytes than to ramified microglia. They do not have the extensive branched appearance of ramified microglia, and their cytoplasm often contains cytoplasmic vacuoles with lipid material. Hickey and Kimura [53] have demonstrated by a chimeric model system that perivascular microglia are derived from blood monocytes. *Ameboid microglia* have a similar morphology as perivascular microglia, and can be unipolar or bipolar. In inflammatory conditions, it is the ameboid microglia that becomes activated and proliferates, ultimately forming microglial nodules as seen in patients with viral encephalitis and ADC [54].

The origin of the microglia is still controversial. Most of the current evidence strongly suggests that microglia arise from mesodermal tissues, and ultimately develop from bone marrow cells, in particular the monocyte [52, 53, 55]. It is thought that the microglia populate the CNS after it has been vascularized. Other groups believe microglia are derived from neuroectoderm, as are macroglia [56].

Microglia Markers

All known phenotypic markers for microglia are shared with other cell types, thus, there are no unique microglia-specific antigens. Microglia can be identified by a number of cell surface antigens which include: immunoglobulin Fc receptors [57], type 3 complement receptors [57, 58] and β_2-integrins [59]. Microglia can also be identified by the presence of the enzyme, nonspecific esterase [60] (fig. 3), and by lectin histochemistry for the lectin *Ricinus communis* agglutinin I (RCA-I) [61]. It should be noted that these antigens/markers are found on other cells of mononuclear phagocyte origin.

Microglia Functions

The major known function of microglia is the phagocytosis of cellular debris, which may be important for tissue modeling in the developing CNS [52]. Also, microglia may be involved with inflammation and repair in the adult CNS due to their phagocytic ability, release of neutral proteinases and production of oxidative radicals. A recent report suggests that microglia processes are incorporated in the layer of astrocytic foot processes of the perivascular glia limitans, and thus may contribute to the integrity of the BBB [62].

As will be discussed later, microglia have been demonstrated to express MHC antigens upon activation, act as APC, secrete a number of immunoregulatory cytokines, and respond to cytokine stimulation, suggesting an involvement with immune responses within the CNS.

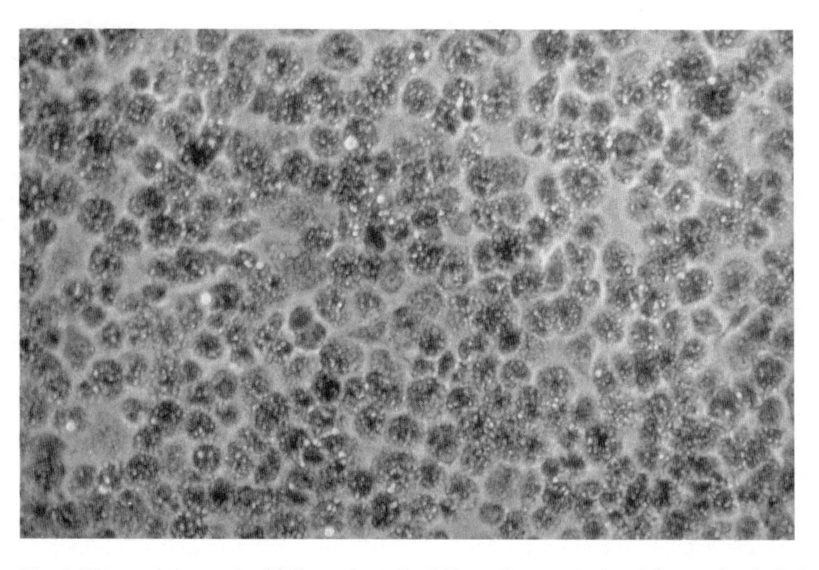

Fig. 3. Photomicrograph of feline microglia. Microglia were isolated from mixed glial cultures on day 10 by mechanical shaking, replated in tissue culture flasks, allowed to adhere for 2 h, and then were stained for the presence of nonspecific esterase. [Picture courtesy of Dr. Edward Hoover and Dr. Steven Dow, Colorado State University.]

Abnormal Glial Cell Function in Neurological Diseases

Although glial cell proliferation is critical in the development of the nervous system, these cells do not normally proliferate in the adult nervous system other than in response to injury, infection or trauma [63, 64]. An early response of the CNS to inflammation is the astrocytic reaction, astrogliosis, which is characterized by proliferation, hypertrophy and the increased synthesis of GFAP. The mechanism(s) by which astrogliosis is induced is poorly understood at this time. Astrogliosis is associated with a number of neurologic disorders including MS, ADC, and EAE. A number of cytokines, including interleukin-1 (IL-1) [65, 66], tumor necrosis factor-α (TNF-α) [67], and interleukin-6 (IL-6) [67], have been implicated in stimulating astrocyte proliferation.

The fate of the myelin-producing cell, the oligodendrocyte, is much less well defined in diseases such as MS and ADC. The oligodendrocyte has a critical role in the disease process of MS, as these cells and their membrane product (myelin) are destroyed, resulting in demyelination. In MS, there initially appears to be a burst of metabolic activity as evidenced by hyperplasia in the margin of active plaques, which is followed by an apparent loss of cells from the center of chronic lesions [63, 64]. There is evidence, however, that oligodendrocytes can survive active lesion formation, and can proliferate under these circumstances [68]. Additionally, it has been documented that remaining oligodendrocytes located at the margins of plaques are able to remyelinate to an extent denuded axons. These 'shadow plaques' are found primarily at the edge of MS lesions, and are composed of thinly remyelinated axons, with the myelin exhibiting aberrant layering [37, 69]. A correlation between oligodendrocyte proliferation and the extent of remyelination has been documented in the animal model of EAE [70], although this proliferative process may be abortive, incomplete, or transient. Oligodendrocytes do not appear to actively participate in immunological events during lesion formation in MS, however, they may be a target cell for a number of cytokines including IL-1 [71], interleukin-2 (IL-2) [72, 73], and TNF-α [74, 75].

In the diseases of MS and EAE, microglia have been shown to phagocytose myelin debris by two pathways: lysosomal degradation and receptor-mediated phagocytosis, which involves IgG as the ligand [76–78]. The disease ADC is due to HIV-1 infection of the CNS. There is overwhelming evidence that perivascular microglia and multinucleated microglia are infected by HIV-1 [79], and a number of cytokines such as IL-1 and TNF-α may activate and enhance HIV replication in these cells [80, 81]. Whether the clinical symptoms and pathology of ADC are linked to altered microglia function within the CNS is not known at this time.

In vivo CNS Sources of Cytokines

What conditions exist in which lymphoid and mononuclear cells would be present in the CNS? Pathological events within the CNS often result in the breakdown of the BBB, which permits cells of the peripheral immune system to enter this site. As mentioned previously, during human diseases such as viral encephalitis [5], MS [6], and ADC [9], and animal models of CNS disease such as EAE [10] and subacute demyelinating encephalomyelitis [82], inflammatory infiltrates comprised of activated T cells, B cells and macrophages are found in the brain. The presence of mononuclear/lymphoid hypercellularity in the area of active plaques, as well as normally myelinated white matter in MS brain, suggests an involvement of the immune system

Table 2. Cytokines present in the CNS during disease states

Cytokine
Interleukin-1
Interleukin-2
Interleukin-6
Interferon-γ
Tumor necrosis factor-α
Lymphotoxin
Transforming growth factor-β

with the disease process of MS, as well as other diseases. The presence of these cells provides the CNS with an endogenous source of cytokines. There is precedence for cytokines to be present in the CNS during neurological disease as: (1) IL-1, interferon-γ (IFN-γ), IL-2, TNF-α and lymphotoxin have been localized in sections of MS brain [83–86]; (2) IL-1 is present in brain from patients with Down's syndrome and Alzheimer's disease [87]; (3) transforming growth factor-β (TGF-β) has been localized in the brains of ADC patients [88]; (4) IFN-γ and IL-6 have been detected in CSF during viral meningitis and encephalitis [89]; (5) elevated levels of IL-6 have been detected in CSF from systemic lupus erythematosis patients having neurological involvement [90]; (6) increased CSF levels of IL-6 and TNF-α occur frequently in MS and in patients with other inflammatory neurological diseases [91]; (7) IL-1 is elevated in CSF from guinea pigs with chronic EAE [92], and (8) increased IL-6 levels have been found in the CNS of mice suffering acute EAE [93] (see table 2 for summary). The presence of these cytokines has been attributed to both the infiltration of activated mononuclear cells, as well as the activation of endogenous glial cells such as the astrocyte and microglia. The situation exists, then, for both cells of the peripheral immune system and glial cells to be in close proximity to each other, and also for immunoregulatory cytokine production by both cell types.

Cytokines

Cytokines play a major role in the initiation, propagation and regulation of immune and inflammatory responses. Cytokines are a diverse group of proteins, with a wide range of functions and target cells. In this section, I will briefly describe the cytokines most relevant to inflammatory and immune responses, especially those involved with these events within the CNS. I will also provide a simplified overview of the components (cells, receptors, cytokines) that are required for the induction of an immune response. There are numerous excellent review articles on cytokine function, so I will reference them for persons needing more detailed information.

Although cytokines comprise a diverse group of proteins, they share a number of general properties. Cytokines are produced during the effector phases of immunity, and serve to mediate and regulate immune and inflammatory responses. Cytokine production is usually transient. Their expression is initiated by activation of gene transcription, the subsequent cytokine mRNA transcripts are unstable, and cytokines are rapidly secreted, resulting in a 'burst' of cytokine release. An individual cytokine can be produced by many different cell types, and have multiple effects on different cell types. Cytokines have also been shown to have redundant functions, i.e. several cytokines can mediate a common event. Thus, the cytokine system displays pleiotropism and redundancy. Cytokines often influence both the synthesis and function of other cytokines, resulting in complex regulatory pathways for immune and inflammatory responses. Cytokines initiate their action by binding to specific cell surface receptors on target cells. These receptors show high affinities for their ligands, with dissociation constants in the range of 10^{-10}–10^{-12} *M*. This suggests that very small amounts of a cytokine need to be produced to elicit a biological response. The ultimate response of target cells to a particular cytokine is determined by the expression of the cytokine receptor, and the nature of the coupling between the receptor and the signal transduction pathways of the target cells.

Tumor Necrosis Factor-α

TNF-α is a 17,000-dalton polypeptide synthesized primarily by activated macrophages during host responses to microbial infection [for review, see 94]. The major cellular source of TNF is the activated macrophage, although many other cell types, including astrocytes and microglia, can be stimulated to secrete TNF-α. TNF-α is the principal mediator of the host response to gram-negative bacteria, and also participates in inflammatory responses. TNF-α can alter vascular endothelial cell function, leading to active participation of this cell type in inflammation. Specifically, TNF-α enhances the permeability of endothelial cells [95], increases expression of adhesion molecules [96], and enhances local adhesion of neutrophils, monocytes and lymphocytes to endothelial cell surfaces [97], thereby facilitating transendothelial migration of immune cells, and establishment of leukocyte-rich inflammatory infiltrates. TNF-α stimulates other cell types to produce cytokines, including IL-1, IL-6, CSFs, and TNF-α itself. TNF-α can also regulate immune responses by modulating the expression of class I and II MHC molecules on a variety of cell types, including astrocytes and microglia. TNF-α is also an endogenous pyrogen which acts on cells in the hypothalamic regions of the brain to induce fever. Long-term systemic administration of TNF-α to animals causes cachexia, a state characterized by wasting of muscle and fat cells. Many of the biological effects of TNF-α are regulated by the cytokine IFN-γ (see description below). Coordinate secretion of these two cytokines may provide a mechanism of locally enhancing the action of TNF-α when low concentrations of TNF-α are produced. cDNA cloning studies have resulted in the identification of two distinct TNF-α receptors, the 55,000-dalton and 75,000-dalton TNF-α receptors [98–100]. The intracellular portions of the two receptors have differences, which suggests that the two receptors may activate different intracellular signaling pathways. The presence of either the 55,000- or 75,000-dalton TNF-α receptor alone is sufficient for high-affinity binding and biological activity.

Interleukin-1

IL-1 is a 17,000-dalton polypeptide produced predominantly by activated macrophages, although other cell types such as endothelial cells, B cells, epithelial cells, keratinocytes, microglia, and astrocytes can also secrete IL-1 upon stimulation [for review, see 101, 102]. IL-1 is a cytokine responsible for mediating a variety of processes in the host response to microbial and inflammatory diseases. There are two forms of IL-1, IL-α and IL-1β, which are the products of two different genes. Although these two forms of IL-1 have less than 30% structural homology to one another, they both bind to the same surface receptor, and have essentially identical biologic activites. When produced locally at low concentrations, the predominant effects of IL-1 are for the most part immunomodulatory. IL-1 is the major co-stimulator for T-cell activation via the augmentation of both IL-2 and IL-2 receptor expression. These effects allow antigen-stimulated T cells to rapidly proliferate and expand in number. IL-1, in cooperation with other cytokines, can enhance the growth and differentiation of B cells. IL-1 is a principal participant in inflammatory reactions through its induction of other inflammatory metabolites such as prostaglandin, collagenase, and phospholipase A_2. In addition, similarly to TNF-α, IL-1 acts on endothelial cells to promote leukocyte adhesion. IL-1 stimulates numerous cell types to produce various cytokines, such as IL-6, TNF-α, CSFs and IL-1 itself. The mouse IL-1 receptor consists of an extracellular IL-1 binding domain (319 amino acids), a membrane-spanning domain (21 amino acids), and a cytoplasmic domain (217 amino acids). A mutated IL-1 receptor which lacks the cytoplasmic domain is able to bind IL-1 with high affinity, but is unable to transmit a biological signal, which indicates that the cytoplasmic domain is critical for signal transduction.

Interleukin-6

IL-6, along with IL-1 and TNF-α, is pleiotropic cytokine involved in the regulation of inflammatory and immunologic responses [for review, see 103–105]. IL-6, a 26,000-dalton molecule, is secreted by a wide range of cells including fibroblasts, monocytes, B cells, endothelial cells, T cells, microglia, and astrocytes. Depending upon the cell type, synthesis of IL-6 is induced by a variety of agents including the cytokines IL-1, TNF-α and IFN-γ. The two best described actions of IL-6 are on hepatocytes and B cells. IL-6 can stimulate hepatocytes to produce several plasma proteins such as fibrinogin and C-reactive protein, which contribute to the acute phase response. IL-6 is the principal cytokine for inducing terminal differentiation of activated B cells into immunoglobulin-secreting plasma cells. A minor function of IL-6 is as a co-stimulator of T cells and of thymocytes. The human IL-6 receptor has an external ligand-binding domain (340 amino acids), and an intracellular domain of 80 amino acids. An anti-IL-6 receptor antibody immunoprecipitates a 130,000-dalton protein (gp 130) that does not bind IL-6 itself, but plays an important role in signal transduction by interacting with the external domain of the IL-6 receptor. IL-1, TNF-α and IL-6 can be grouped as cytokines mediating common biological effects.

Interleukin-2

IL-2, a 14,000 to 17,000-dalton glycoprotein, is the principal cytokine responsible for proliferation and maintenance of T cells [for review, see 106]. IL-2 is produced primarily by $CD4^+$ T-helper cells when they are activated, and acts in both an autocrine and paracrine fashion on T cells. IL-2 synthesis is transient, with an early peak of secretion occurring approximately 4 h after stimulation. The principal actions of IL-2 are on lymphocytes, and include functioning as the major autocrine growth factor for T cells. Binding of IL-2 by T cells results in proliferation of these cells, enhanced secretion of other cytokines such as IFN-γ and lymphotoxin, and the enhanced expression of transferrin receptors. IL-2 can also stimulate the growth of NK cells, activated B cells, and macrophages.

T cells express two distinct surface proteins which bind IL-2. A 55,000-dalton protein (α subunit; p55) is expressed when T cells are activated; p55 binds IL-2 with low affinity (K_d of approximately 10^{-8} *M*). Binding of IL-2 to T cells which express only p55 generally does not lead to any biological response. The second IL-2-binding protein has a molecular weight of 75,000 daltons (β subunit; p75) and binds IL-2 with an intermediate affinity of 10^{-9} *M*. Cells that express both p55 and p75 bind IL-2 with a high affinity of approximately 10^{-11} *M*. It has been proposed that p55 forms a complex with p75, which increases the affinity of the p75 receptor for IL-2, thereby rendering T cells responsive to low IL-2 concentrations.

Interferon-γ

IFN-γ, or immune interferon (MW of 17,000 daltons), is produced predominantly by activated $CD4^+$ and $CD8^+$ T cells, although NK cells are also capable of producing IFN-γ [for review, see 107]. IFN-γ is a pleiotropic cytokine with antiviral activity, antiproliferative effects, and immunomodulatory effects. These immune effects include activation of mononuclear phagocytes; enhancing the generation of toxic oxygen radicals by macrophages; modulating both class I and II MHC expression on epithelial cells, endothelial cells, macrophages, and glial cells; promoting differentiation of both T and B cells, and activating neutrophils, NK cells and vascular endothelial cells. As mentioned previously, IFN-γ can also enhance the action of TNF-α, in part, by increasing TNF-α receptor expression. The IFN-γ receptor has an extracellular domain, membrane-spanning domain, and cytoplasmic domain. However, the IFN-γ receptor protein itself cannot transduce a biological signal, suggesting that additional components are required for signal transduction.

Components of the Immune System

Cells of the immune system mediate multiple processes including elimination of foreign pathogens, neutralization of toxins, and killing of tumor cells. The hallmark of an immune response is *antigen specificity,* which lies in the specific cellular recognition of foreign antigens. This specificity is conferred by the two major classes of lymphocytes, T cells and B cells, which contain in their unique surface receptors the molecular basis for antigenic specificity.

Cells of the immune system originate from pluripotent hematopoietic stem cells in the bone marrow which give rise to two different cell lineages: the lymphoid cells, which are the precursors of B and T cells, and the myeloid cells, which give rise to monocytes (macrophages), neutrophils, eosinophils, erythrocytes, and mast cells.

The two classes of lymphocytes, T cells and B cells, differ in their functional properties. T cells are responsible for cell-mediated immunity, as well as coordinating the functions of various other cell types, including B cells. T cells can be broadly divided into two subsets which both have antigen-specific T-cell receptors (TCR) on their surface. T cytotoxic cells, identified by the surface molecule CD8, can destroy virus-infected cells or tumor cells. The TCR on $CD8^+$ cells recognizes foreign antigen in conjunction with class I MHC. The second type of T cell, T-helper cells, regulate the ability of B cells to produce and secrete immunoglobulin, and are critical for the initiation of immune responses. T-helper cells, identified by the surface molecule CD4, recognize antigen in association with class II MHC molecules. B cells are primarily effector cells. They secrete antigen-specific immunoglobulin molecules that mediate humoral immunity. B cells also express cell surface immunoglobulins that function as specific antigen receptors.

Macrophages are a third class of cells (accessory cells) that do not possess any specific antigen-recognition capacity, but are critical for the functioning of the immune system. Macrophages play an essential role in the process of antigen presentation to T cells by allowing T-cell recognition of foreign antigens, and also provide to both T cells and B cells the extracellular signals (cytokines) required for their functional activation. Macrophages are considered the traditional APC, although a number of cell types, including astrocytes and microglia, can acquire this property.

Cooperation between the different lymphoid cells (T and B cells) and macrophages (or other APC) is required for the initiation, perpetuation, and ultimate down-regulation of an immune response (fig. 4). The initial activation of T-helper cells depends upon the ability of the T-helper cell (via its antigen-specific TCR) to recognize foreign antigen in association with class II MHC on an APC. Cytokines such as IL-1 are also produced by APC, and are required for co-stimulation of T-helper cells. Once activated, T-helper cells proliferate and expand. These T cells then help B cells to proliferate, differentiate, and ultimately produce immunoglobulin. Antigen-specific T-cell help to B cells may be delivered by cytokines or direct cellular interactions. Activated T-helper cells also release cytokines such as IFN-γ which

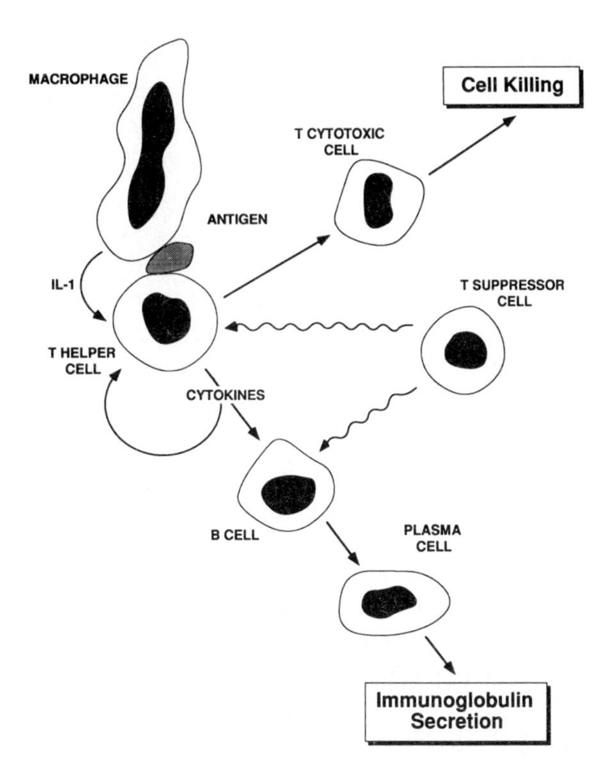

Fig. 4. Interaction of macrophage and lymphoid cells for induction of an immune response. Macrophage (APC) presents foreign antigen in conjunction with class II MHC molecules to $CD4^+$ T-helper cells. T-helper cells then become activated, secrete cytokines, and participate in helping B cells differentiate into plasma cells (humoral immunity), and regulating T-cell effector functions (T cytotoxic cells, T suppressor cells) (cell-mediated immunity).

activates macrophages, and cooperate with T cytotoxic cells in their recognition of virally infected cells or allogeneic grafts. T suppressor cells act as regulatory cells to suppress ongoing immune responses.

Immune Cell-Derived Cytokines and Their Effect on Glial Cells

Studies in this field were initially directed towards investigating whether factors from immune cells may be contributing to astrocytic activation seen in the region of inflammatory infiltrates in diseases such as MS. Fontana et al.

[108] tested this hypothesis by examining the ability of lymphocytes to produce glial-stimulating factors in vitro. Supernatants from rat lymphocytes stimulated with the mitogen concanavalin A or sensitized rat lymphocytes challenged with antigen (bovine γ-globulin) in vitro, were collected and tested for glial-stimulating activity. Indeed, they found that these supernatants enhanced both RNA synthesis and DNA synthesis in rat astrocyte cultures [108]. This factor, named glial-stimulating factor (GSF), was produced by both activated T and B lymphocytes. A factor with similar activity was also derived from mitogen-activated human T lymphocytes [109]. Human GSF activity was detected in two molecular weight peaks, one at 30,000 daltons and the other at less than 10,000 daltons [109]. The 30,000-dalton GSF activity appeared to co-migrate with IL-1 activity. Additionally, the T cell line MOLT-4 and the B cell line RPMI-1788 spontaneously secreted GSF activity, but those factors were not subjected to molecular weight determination.

Merrill et al. [110] demonstrated that both rat astrocytes and oligodendrocytes could respond by increased proliferation to supernatants from activated human T cells. The supernatants tested were derived from a T lymphoblast line (MO) which was infected with HTLV-II, human T lymphocytes transformed by HTLV-II, and human T lymphocytes activated by the mitogen phytohemagglutinin. Additionally, the MO cell line also produced factors which enhanced maturation of oligodendrocytes as assessed by increased expression of MBP and GPDH. The above studies indicate that lymphoid cells activated by three different mechanisms: (1) nonspecifically by mitogen; (2) in an antigen-specific fashion, and (3) by viral transformation, are all capable of secreting factors which stimulate the growth of glial cells.

A growth factor specific for oligodendrocytes has been purified from the MO T cell line [111]. This factor, termed glial growth promoting factor (GGPF), is constitutively produced by the MO T cell line, and has a molecular weight of 30,000. GGPF stimulates the proliferation of oligodendrocytes, and has no apparent stimulatory activity on rat astrocytes. Biochemical and biological characterization of GGPF suggests that it is a functionally distinct lymphokine from those previously described.

The availability of recombinant cytokines has greatly facilitated the study of cytokine influences on glial cell function, as the use of crude cell-derived supernatants is always complicated by the fact that numerous cytokines can be present in these preparations. In the next sections, I will summarize what is known about the effects of recombinant cytokines such as IL-1, IL-2, IFN-γ, TNF-α, and IL-6 on glial cell function.

Interleukin-1 and Glial Cells

Biological Effects of IL-1 on Glial Cells: Astrocytes

As mentioned earlier, IL-1 affects a wide range of target cells, and is considered an important mediator of inflammatory responses. IL-1 may also mediate inflammation associated with injury to the brain. As one response to brain injury is proliferation of astrocytes, IL-1 was tested for its capacity to induce proliferation of these cells. Purified IL-1 was shown to have a stimulatory activity for astrocyte growth in vitro [65], while IL-1 injected into brain can stimulate astrogliosis [66]. These results suggest that IL-1, released by inflammatory cells, may contribute to astroglial scarring in damaged mammalian brain. As will be discussed in more detail later, both astrocytes and microglia, upon stimulation, secrete an IL-1-like molecule. This would provide an endogenous brain source of IL-1, which could promote astrogliosis and the development of immune responses within the CNS. Additionally, IL-1 may be acting as an autocrine stimulator of astrocytes, as these cells both secrete and respond to this cytokine. Recombinant IL-1 has also been shown to stimulate the proliferation of a human astrocytoma cell line, U373 [112]. This indicates that not only do cultures of rat astrocytes proliferate in response to IL-1, but a human GFAP-positive astrocytoma cell line does as well.

IL-1 has also been demonstrated to modulate gene expression in astrocytes. IL-1 can increase the expression of ICAM-1 adhesion molecules on human astrocytes [113]. The expression of adhesion molecules is thought to facilitate the ability of the astrocyte to function as an APC in the CNS, and influence intracerebral immune responses. This is particularly important as the expression of ICAM can enhance the ability of an APC to present antigen when the number of MHC molecules on the cell surface is low, suggesting that the combined expression of ICAM and MHC may be synergistic in eliciting a more potent immune response [114].

IL-1 is also a potent inducer of cytokine production by astrocytes. IL-1 stimulation of primary rat astrocytes results in the secretion of TNF-α [115] and IL-6 [116, 117], while human astroglial cell lines will produce CSFs [118], TNF-α [119], and IL-6 [120] in response to IL-1. These cytokines (IL-1, TNF-α, IL-6) have wide ranging effects on glial cells themselves, which are summarized in tables 3, 6 and 7. Little is known about the biological effect of IL-1 on microglia and oligodendrocytes, although microglia are capable of secreting IL-1 [121].

Table 3. IL-1 and glial cells

Astrocytes
Induces proliferation of neonatal astrocytes, human astroglioma cells
Induces ICAM-1 expression on fetal human astrocytes
Induces IL-6 expression by neonatal astrocytes, human astroglioma cell lines
Induces TNF-α expression in conjunction with IFN-γ by neonatal rat astrocytes
Induces TNF-α expression in human astroglioma cell lines
Induces G-CSF and GM-CSF expression in human astroglioma cell lines
Astrocytes make IL-1 in response to LPS stimulation
Human glioblastoma cells constitutively secrete IL-1
Microglia
Microglia make IL-1 in response to LPS stimulation

IL-1 Production by Glial Cells

Studies were initiated to determine whether glial cells might secrete soluble factors that would regulate lymphoid/mononuclear cells, or glial cells themselves. Fontana et al. [122] discovered that cultured murine astrocytes, upon stimulation with LPS, secreted significant amounts of prostaglandin and an IL-1-like factor. These two immunoregulatory molecules are also secreted by the C6 glioma cell line, which is thought to represent undifferentiated astrocytic cells [122]. C6 cells spontaneously secrete PGE and IL-1, and the IL-1-like factor was increased upon incubation with LPS and indomethacin. Human glioblastoma cell lines also constitutively secrete an IL-1-like molecule [123]. Müller cells, the major glial cell of the retina, have been shown to secrete IL-1 upon stimulation with LPS or IFN-γ [124]. Both rat and murine microglia are also capable of producing IL-1 in response to LPS stimulation [121, 125], and virally transformed microglia clones produce IL-1 [126].

These findings indicate that there are two endogenous sources of IL-1 within the CNS: the astrocyte and microglia. Giulian et al. [66, 121] propose that microglia are the more likely source of IL-1 during acute phases of brain injury. Since microglia are the first brain cells to appear in increased numbers at sites of trauma or infection, it has been suggested that IL-1 released from microglia stimulates the proliferation of nearby astrocytes. As astrocytes also produce IL-1, this type of circuitry provides a potential autocrine mechanism for astrocytic growth, which may be critical in the promotion of astroglial scarring in diseased or traumatized brain. Table 3 summarizes the effects of IL-1 on glial cells.

Role of IL-1 in EAE

A variety of animal models exist for the study of CNS disease involving inflammatory demyelinating lesions. The best characterized experimental model for CNS autoimmune disease is EAE. This disease is induced by injection of MBP or transfer of encephalitogenic MBP-specific T cells to naive recipients. EAE is characterized by invasion of the CNS by T lymphocytes and macrophages, demyelination, and acute, chronic or chronic relapsing paralysis. The mediators of this disease are MBP-reactive T-helper cells which are class II MHC restricted [for review article, see 127]. It has been suggested that antigen-specific autoimmune T cells are responsible for initiation of disease, and that perpetuation of disease and subsequent demyelination may be the result of an influx of largely nonantigen-specific inflammatory cells of the recipient animal [128]. There appears to be genetic control of susceptibility to EAE. In inbred rat strains, Lewis rats (RT-1^{l} haplotype) are susceptible to EAE, whereas Brown-Norway (BN) rats (RT-1^{n}) are resistant. Disease susceptibility appears to be linked to MHC alleles, although non-MHC genes may play a small role in contributing to EAE [129–131].

Due to the inflammatory effects of IL-1, and the knowledge that glial cells can produce IL-1 within the CNS, there has been interest in the role of IL-1 in the development of EAE. IL-1 has been shown to enhance the in vitro activation of encephalitogenic T cells, thereby enhancing the adoptive transfer of EAE [132]. More recently, Jacobs et al. [133] demonstrated that in vivo administration of IL-1α enhanced the severity and chronicity of clinical paralysis associated with EAE, while treatment of animals with soluble mouse IL-1 receptor (an IL-1 antagonist) significantly delayed the onset of EAE, reduced the severity of paralysis and weight loss, and shortened the duration of disease. These results implicate the involvement of IL-1 in inflammatory CNS disease states. As IL-1 does have varied effects on endogenous glial cells such as expression of adhesion molecules, cytokine induction, and astrogliosis, the inhibition of IL-1 activity by soluble IL-1 receptor may prevent further CNS inflammation and injury. It is speculated that the use of soluble IL-1 receptors as an IL-1 antagonist may be therapeutically beneficial in diseases like MS.

Interleukin-2 and Glial Cells

Biological Effects of IL-2 on Glial Cells: Oligodendrocytes

IL-2 is the product of activated T cells, thus, would only be present in the CNS during pathological disease states associated with lymphoid infiltration.

Recombinant human IL-2 has been shown to influence the proliferation and differentiation of oligodendrocytes [72]. GalC-positive oligodendrocytes are increased approximately 3-fold in cultures containing IL-2. Additionally, IL-2 appears to stimulate the maturation of oligodendrocytes, assessed as the enhanced expression of MBP in IL-2-treated oligodendrocytes. While most oligodendrocytes express the surface antigen GalC [42], the simultaneous expression of GalC and MBP implies a more differentiated state [45]. Both MBP mRNA and protein levels are increased in IL-2-stimulated oligodendrocytes [134]. MBP mRNA levels increase within 8 h after IL-2 stimulation, while MBP protein levels increase 24 h after stimulation. These findings provide evidence that an important step in oligodendrocyte differentiation – MBP mRNA and protein expression – can be amplified in part by IL-2. Two human glioblastoma cell lines with oligodendroglial phenotype ($GalC^+$, $GFAP^-$) proliferate in the presence of IL-2, and also bear receptors for the IL-2 molecule as determined by immunofluorescent staining with anti-IL-2 receptor monoclonal antibodies [135]. These findings indicate that both $GalC^+$ primary oligodendrocytes and $GalC^+$ glioblastoma cell lines can respond in a mitogenic fashion to IL-2. Human oligodendroglioma cell lines expressing only IL-2 p75 receptors will proliferate in response to IL-2, suggesting that the presence of p75 itself is sufficient to deliver the IL-2 signal to these cells [136]. Interestingly, IL-2 appears to have an opposite effect on oligodendrocyte progenitor cells [73]. Human IL-2 (purified, not recombinant) inhibited the proliferation of $A2B5^+$ rat oligodendrocyte progenitor cells, and this inhibition was dependent on the expression of Tac receptors on these cells (now known as p55 receptors). These results suggest that IL-2 can have varying biological effects on oligodendrocytes depending on the maturity of those cells, i.e. $GalC^+$, $A2B5^-$ oligodendrocytes respond in a positive manner to IL-2 (enhanced proliferation and maturation), while $GalC^-$, $A2B5^+$ progenitor cells respond in a negative fashion (inhibition of proliferation). See table 4 for a summary of the effects of IL-2 on glial cells. IL-2 has not been shown to modulate the function or gene expression of astrocytes or microglia, thus, it appears to have a selective effect on oligodendrocytes.

Interferon-γ and Glial Cells

Biological Effects of IFN-γ on Glial Cells

Class I MHC Expression

As mentioned previously, IFN-γ is the product of activated T cells, and has a wide range of immunoregulatory functions. IFN-γ would thus be

Table 4. IL-2 and glial cells

Oligodendrocytes
Induces proliferation of rat oligodendrocytes, human oligodendroglioma cell lines
Enhances expression of MBP in rat oligodendrocytes
Inhibits proliferation of oligodendrocyte progenitor cells

present in the CNS only during disease states where the BBB has been breached, and activated T cells have infiltrated into that site. IFN-γ has been shown to interact with astrocytes, oligodendrocytes and microglia, and functions in part by modulating MHC gene expression on these cells. Cells in the brain express extremely low levels of antigens encoded by the MHC. These antigens, class I and class II, play critical roles in the initiation and propagation of immune responses. Brain cells such as astrocytes, oligodendrocytes, and microglia have been shown to constitutively express low levels of class I antigens [2]. IFN-γ induces a substantial increase in the expression of class I antigens on all these cell types, in a wide variety of species (rat, mouse, human) [2, 137, 138]. The direct injection of IFN-γ into the brains of mice also enhances class I MHC expression in vivo [2]. Class I antigens are involved in antigen recognition by cytotoxic T cells, which usually results in destruction of class I-positive cell targets [139]. The implication of enhanced class I MHC antigen expression on glial cells by IFN-γ is that they can be rendered susceptible to lysis by class I-restricted cytotoxic T cells. Astrocytes in vitro can serve as targets for class I MHC-specific cytotoxicity mediated by CTL [140].

In vitro Class II MHC Expression

Astrocytes and microglia do not constitutively express class II MHC antigens, however, IFN-γ can induce the expression of class II MHC molecules on these cells [141–144]. Class II MHC expression has not been detected on oligodendrocytes, regardless of any stimuli tested. The normal function of class II antigens is their involvement in antigen presentation by accessory cells to T helper cells. The accessory cell activity of macrophages, i.e. presentation of antigen, has been demonstrated to be proportional to the amount of class II MHC induced on these cells following exposure to IFN-γ. Modulation of class II expression by IFN-γ may provide a regulatory mechanism for local immune responses. Fontana et al. [142] have demonstrated that astrocytes are able to present antigen to T cells in an MHC-restricted manner. Specifically, astrocytes expressing class II antigens can present MBP to MBP-specific T-cell lines, resulting in T-cell activation. The implication of class II expression on astrocytes, and their antigen-presenting capability, is that these cells may stimulate the development of aberrant immune responses within the CNS, possibly autoimmune reactions. This hypothesis, however, requires the presence of activated, IFN-γ-secreting T cells within the CNS. Massa et al. [145, 146] have shown that viral infection of astrocytes (hepatitis virus and measles virus) also results in the expression of class II antigens. This provides a mechanism for astrocyte class II expression independent, at least initially, of activated T cells and IFN-γ secretion. It has been proposed that both brain endothelium and astrocytes are involved in CNS immune responses. In light of the fact that lymphocyte/monocyte traffic through the brain is normally limited, the cerebral endothelium, which in part maintains the BBB, is thought to be important in the induction of immune responses, since lymphocytes may first encounter brain antigens on these cells. In fact, endothelial cells can express class II antigens upon treatment with IFN-γ [147, 148] and present antigen [148, 149]. Class II MHC-positive endothelial cells could then act as the first APC encountered by blood-borne T cells, resulting in activation of these cells. These activated T cells would then be able to penetrate the BBB in a variety of fashions such as passing through altered tight junctions between the endothelial cells, or by dissolving the extracellular matrix produced by endothelial cells. Activated T cells have been shown to secrete an endoglycosidase that digests the proteoglycans of the extracellular matrix [150]. Once activated T cells have penetrated the BBB, the first cells they would encounter are astrocytes. The secretion of IFN-γ in such close proximity to astrocytes could then act as a signal for class II MHC expression on these cells. Class II-positive astrocytes could then act as APC leading to further activation of T lymphocytes, and, in concert with IL-1 production by astrocytes and microglia, create an environment for local immune responses within the CNS.

Recent studies from our laboratory have focused on understanding the intracellular signaling events involved in IFN-γ induction of class II MHC expression by astrocytes. Our results indicate that both IFN-γ-induced activation of protein kinase C (PKC) and enhanced Na^+ influx are required for class II MHC gene expression in the astrocyte [151]. The PKC activator, phorbol myristate acetate (PMA), was not able to mimic the action of IFN-γ, indicating that activation of PKC alone is not sufficient to induce class II MHC expression. In the rat astrocyte, both IFN-γ-induced PKC activation and Na^+ influx contribute to class II gene expression, although other intracellular signals induced by IFN-γ must be required for the maximal response to this cytokine. In this regard, preliminary data from our laboratory indicates that an IFN-γ-induced nuclear protein is critical for transcription of the class II MHC gene in astrocytes, and that PMA is not capable of inducing this protein [Panek and Benveniste, unpubl. observation]. These results suggest that PKC activation alone by PMA is not sufficient to induce this nuclear protein, and that IFN-γ-induced intracellular signals are more complex than those induced by PMA.

In vivo Class II MHC Expression

Although class II expression on astrocytes has been conclusively demonstrated in vitro, in vivo studies have generated conflicting results. Direct injection of IFN-γ into the brains of mice induced class II antigens on astrocytes, indicating that astrocytes have the potential to express these antigens in vivo [2]. Many laboratories have examined whether astrocytes express class II antigens in a variety of immune-mediated disease states to better understand the possible role of the astrocyte as a local APC. Traugott et al. [152] demonstrated class II expression on astrocytes in active chronic MS lesions, and then confirmed these studies by performing double staining for both class II and GFAP [85]. A study by Hofman et al. [83] also identified class II-positive astrocytes in MS brain by double staining. Immunocytochemistry and electron microscopic analysis of CNS tissue from 3 patients with acute and chronic progressive MS indicate that some astrocytes are class II MHC-positive, and can act as phagocytic cells in the CNS, providing evidence that the astrocyte may indeed function as an APC [153]. Rodriguez et al. [154] have studied class II expression on glial cells in an animal model of CNS demyelination induced by Theiler's virus. In susceptible strains of mice (BIO.S and BIO.ASR2), the majority of class II-positive glial cells had morphological characteristics of astrocytes, while uninfected mice or resistant strains ((BI).S(9R)) were class II-negative. In SJL mice with acute or chronic relapsing EAE, some class II-positive cells were identified as astrocytes [155]; however, other studies investigating the EAE model in Lewis rats failed to detect class II-positive astrocytes in the brain [156–158]. These conflicting results may be due solely to technical problems involved with antigen fixation and staining methodologies, or may indicate that the ability of astrocytes to function as APC in vivo may only be relevant in certain diseases or disease models. Another possibility may be the loss of class II-positive astrocytes by class II MHC-restricted T-cell mediated cytotoxicity as shown by Sun and Wekerle [159].

Another potential explanation for the paucity of class II-positive astrocytes in disease states compared to class II-positive microglia is that class II expression on astrocytes is more readily subject to down-regulation. Sasaki et al. [160] demonstrated that increases in cAMP levels will inhibit astrocyte, but not microglia, class II expression. Norepinephrine has been shown to inhibit astrocyte class II MHC via β_2-adrenergic signal transduction mechanisms, by increasing cAMP levels [161]. Thus, expression of class II on astrocytes may be more susceptible to down-regulation by endogenous agents like norepinephrine, and as such, expression may be more transient on the astrocyte compared to microglia.

Expression of Class II Antigens on Astrocytes: Correlation with EAE

EAE appears to be strain-specific as BN rats and BALB/c or C57BL/6 mice are resistant, whereas Lewis rats and SJL mice are susceptible. Several studies have examined what might contribute to the immunopathological reaction seen in the CNS of susceptible animals. Massa et al. [162, 163] demonstrated that astrocytes derived from susceptible strains express much higher levels of class II antigens upon treatment with either IFN-γ or virus compared to astrocytes prepared from EAE-resistant strains. This hyperinduction of class II in EAE-susceptible animals was astrocyte-specific as both peritoneal macrophages and microglial cells of susceptible and resistant strains showed identical patterns for class II induction. This differential expression of class II on astrocytes compared to microglia suggests that regulation of class II expression on astrocytes may correlate with antigen-presenting capacity and, ultimately, disease development in the CNS.

Modulation of Astrocyte Gene Expression by IFN-γ

IFN-γ can also increase the expression of ICAM-1 adhesion molecules on human astrocytes, in a similar fashion to IL-1 enhancement of ICAM-1 [113]. Primary rat astrocytes express one class of high-affinity TNF-α receptors, which are increased in number upon exposure to IFN-γ [164]. As IFN-γ and TNF-α often synergize for mediating biological effects, the ability of IFN-γ to increase TNF-α receptor expression may contribute, in part, to the synergy observed between these two cytokines. The human astroglioma cell line D54-MG expresses two classes of TNF-α receptors; intermediate and high affinity [165]. IFN-γ enhances the expression of both classes of TNF-α receptors on these cells. For both primary rat astrocytes and human astroglioma cells, IFN-γ increases the number of TNF-α receptors, but has no effect on the binding affinity of the receptors.

IFN-γ does not appear to directly induce cytokine production by astrocytes, but provides a 'priming signal' which renders the astrocyte responsive to a subsequent exposure to other cytokines. For example, neither IFN-γ nor IL-1β alone induce TNF-α production by primary rat astrocytes, but act together in a synergistic fashion to induce TNF-α expression [115]. More importantly, astrocytes pretreated with IFN-γ, then exposed to IL-1β, produce more TNF-α compared to the simultaneous addition of both cytokines, whereas cells pretreated with IL-1β, then exposed to IFN-γ, produce negligible levels of TNF-α. This suggests that IFN-γ generates a priming signal for the astrocytes which then increases their sensitivity to a subsequent exposure to IL-1β. IFN-γ alone has no effect on IL-6 production by astrocytes, but enhances the ability of IL-1β to induce IL-6 expression [117]. The nature of the IFN-γ-induced priming signal(s) is unknown at this time.

Components of the complement cascade have been implicated in contributing to the pathology of several neurological autoimmune diseases such as MS, EAE and Guillain-Barré syndrome [for review, see 166]. Recent studies have indicated that astrocytes can serve as a local endogenous source of some of the complement components, notably C3, the central component of the complement cascade [167]. In addition, IFN-γ can enhance C3 expression in both human astroglioma cells and primary rat astrocytes [168]. This is particularly interesting as IFN-γ either inhibits or has no effect on C3 expression in other cell types such as hepatocytes, monocytes and endothelial cells. Thus, the IFN-γ-mediated increase in C3 gene expression may be unique to the astrocyte. The effects of IFN-γ on glial cells are summarized in table 5.

Role of IFN-γ in EAE

IFN-γ has been demonstrated to be involved in MS lesion exacerbation and pathogenesis [169]. Due to its ability to modulate expression of MHC antigens and adhesion molecules on glial cells, and activate monocytes and macrophages, IFN-γ has been proposed to contribute to the pathogenesis of MS and other neural autoimmune diseases. Curiously, in the animal model of EAE, treatment of mice with neutralizing monoclonal antibody against IFN-γ caused an increase in morbidity rates and mortality, while treatment with IFN-γ itself resulted in reduced morbidity and mortality [170]. These results suggest that both endogenous as well as exogenous IFN-γ exert an

Table 5. IFN-γ and glial cells

Astrocytes
Increases class I MHC expression on primary astrocytes
Induces class II MHC expression on primary astrocytes, human glioma cells
Induces ICAM-1 expression on fetal human astrocytes
Increases TNF-α receptor expression on primary astrocytes, human glioma cells
Primes rat astrocytes for TNF-α production
Primes rat astrocytes for IL-6 production
Enhances expression of the complement component C3 in primary astrocytes, human glioma cells
Microglia
Increases class I MHC expression
Increases class II MHC expression
Induces TNF-α expression
Oligodendrocytes
Increases class I MHC expression

inhibitory effect on the development of EAE. These findings were rather surprising as it was expected that IFN-γ would promote EAE disease progression rather than lessen it. The cause of the contradictory actions of IFN-γ in EAE compared to MS are currently under investigation.

Tumor Necrosis Factor-α and Glial Cells

Biological Effects of TNF-α on Glial Cells

TNF-α is a pleiotropic cytokine synthesized by a variety of cell types, and is recognized to be an important mediator of inflammatory responses in a variety of tissues, and also to function as an immunoregulatory cytokine. TNF-α has a diverse range of functions in the CNS due to its influence on astrocytes and oligodendrocytes, which will be reviewed in this section.

TNF-α has been shown to mediate myelin and oligodendrocyte damage in vitro [75], and has cytotoxic activity against rat oligodendrocytes, which results in cell death [74]. This aspect of TNF-α activity could certainly contribute to myelin damage and/or the demyelination process seen in diseases such as MS, EAE and ADC. TNF-α also has multiple effects on the astrocyte which are noncytotoxic in nature, and may function in an autocrine manner as astrocytes express specific high affinity TNF-α receptors [164], and secrete TNF-α upon activation by a variety of stimuli [115, 171]. TNF-α increases class I MHC expression on astrocytes [172], thereby making the astrocyte a more susceptible target for class I-restricted CTL [140]. TNF-α alone has no influence on astrocyte class II expression, but acts to enhance class II MHC expression induced by IFN-γ or virus [164, 173]. TNF-α acts by increasing IFN-γ-induced transcription of the class II gene, rather than having an effect on class II mRNA stability [174]. TNF-α also induces ICAM-1 expression on astrocytes [113]. The effect of TNF-α on both class II MHC and ICAM-1 expression would contribute to the ability of the astrocyte to function as an APC within the CNS. TNF-α has a mitogenic effect on both primary astrocytes [67] and human astroglioma cell lines [112, 165], which is thought to contribute to the reactive astrogliosis associated with various neurological diseases.

TNF-α also has a multitude of effects on endothelial cells, which function as an interface between blood and surrounding tissues, and are critical for maintaining homeostasis. TNF-α can alter endothelial cell function, leading to the active participation of this cell type in inflammatory reactions. Specifically, TNF-α enhances the permeability of endothelial cells [95], increases ICAM-1 expression [96], induces IL-1 and IL-6 synthesis by endothelial cells [175], and enhances local adhesion of lymphocytes and monocytes to endothelial cell surfaces [97]. The architecture of the BBB, with astrocytic processes abutting onto cerebrovascular endothelium, suggests that astrocyte-derived TNF-α can influence neighboring endothelial cells, alter BBB permeability, and promote inflammatory infiltration into the CNS.

TNF-α induces cytokine production by astrocytes. Astrocytes respond to TNF-α by secreting IL-6 [116, 117], a pleiotropic cytokine involved in B-cell differentiation and immunoglobulin synthesis [105]. Astrocytes also produce G-CSF and GM-CSF in response to TNF-α [125]. G-CSF/GM-CSF can augment inflammatory responses by attracting granulocytes and macrophages to migrate to inflammatory sites in the CNS, by promoting their survival in these sites, and by increasing their effector function. Additionally, GM-CSF can induce the proliferation and activation of microglia [176]. Finally, astrocytes stimulated with TNF-α express TNF-α mRNA, suggesting a positive feedback loop for TNF-α expression [Chung and Benveniste, unpubl. observation].

TNF-α Production by Glial Cells

There are two endogenous sources of TNF-α within the CNS: the astrocyte and microglia. There appear to be multiple stimuli by which TNF-α

Table 6. TNF-α and glial cells

Astrocytes
Increases class I MHC expression on primary astrocytes
Enhances class II MHC expression induced by IFN-γ or virus on primary astrocytes
Induces ICAM-1 expression on human fetal astrocytes
Induces proliferation of adult astrocytes, human astroglioma cell lines
Induces IL-6 production by primary astrocytes
Induces G-CSF and GM-CSF production by primary astrocytes
Astrocytes make TNF-α in response to LPS, virus, IFN-γ/IL-1
Human astroglioma cells make TNF-α in response to IL-1, PMA, calcium ionophore
Microglia
Microglia make TNF-α in response to LPS or IFN-γ
Oligodendrocytes
Cell death
Myelin damage

production is induced in astrocytes. These include: (1) treatment with LPS [74, 115, 171]; (2) exposure to the cytokines IFN-γ and IL-1β [115, 119]; (3) exposure to the neurotropic paramyxovirus, Newcastle disease virus [171], and (4) treatment with phorbol ester and calcium ionophore [165]. Mouse microglia produce TNF-α in response to LPS or IFN-γ [176]. Although numerous published reports have described astrocyte TNF-α production, some controversy remains as to whether the astrocyte is the true producer of TNF-α. The possibility exists that residual contaminating microglia, which also secrete TNF-α upon stimulation, are actually the source of TNF-α in these cultures. The findings that human astroglioma cell lines express mRNA for TNF-α and synthesize biologically active TNF-α upon stimulation implicate the astrocyte as a source for TNF-α [119, 165]. See table 6 for a summary of the effects of TNF-α on glial cells.

Expression of TNF-α by Astrocytes: Correlation with EAE

Cytokine production has been implicated in contributing to autoimmune diseases [for review article, see 177]. Since cytokines play a major role in regulating immune responses, aberrant expression may be a factor in the initiation and perpetuation of autoimmunity. Of particular interest is the fact that the genes for TNF-α and functionally related TNF-β (lymphotoxin) map within the MHC gene complex [178]. Since many autoimmune diseases such as EAE are strongly associated with class II MHC gene products, TNF-α/TNF-β are plausible candidates for cytokines involved with autoimmunity.

We have recently demonstrated that astrocytes from EAE-susceptible and resistant rat strains differ in their ability to express TNF-α mRNA and protein [179]. Astrocytes from EAE-resistant BN rats express TNF-α mRNA and protein in response to LPS alone, yet IFN-γ does not significantly enhance LPS-induced TNF-α expression, nor do they express appreciable TNF-α in response to the combined stimuli of IFN-γ/IL-1β. In contrast, astrocytes from Lewis rats (EAE-susceptible) express low levels of TNF-α mRNA and protein in response to LPS, and are extremely responsive to the priming effect of IFN-γ for subsequent TNF-α gene expression. Also, Lewis astrocytes produce TNF-α in response to IFN-γ/IL-1β. The differential TNF-α production by astrocytes from BN and Lewis strains is not due to the suppressive effect of prostaglandins because the addition of indomethacin does not alter the differential pattern of TNF-α expression. Furthermore, Lewis and BN astrocytes produce another cytokine, IL-6, in response to LPS, IFN-γ and IL-1β in a comparable fashion. Peritoneal macrophages and neonatal microglia from Lewis and BN rats are responsive to both LPS and IFN-γ priming signals for subsequent TNF-α production, suggesting that differential TNF-α expression by the astrocyte is cell-type specific. The capacity for TNF-α production by Lewis astrocytes, especially in response to disease-related cytokines such as IFN-γ and IL-1β, may contribute to disease susceptibility and to the inflammation and demyelination associated with EAE. We are currently investigating the molecular mechanisms underlying differential astrocyte TNF-α gene expression in these two rat strains.

Role of TNF-α in EAE

An association of TNF-α with the disease of MS is suggested by the observations of TNF-α-positive astrocytes in MS brain [84, 86]. Circumstantial evidence for the role of TNF-α/TNF-β in EAE was obtained from studies demonstrating that the ability of encephalitogenic T cells clones to transfer disease was positively correlated with the amount of TNF-α/TNF-β cytotoxic activity [180]. More conclusive evidence was recently obtained by Ruddle et al. [181], in which they demonstrated that an antibody to TNF-α/TNF-β could prevent the transfer of EAE by encephalitogenic T cells. These findings indicate that inhibition of the biological activities of these two cytokines can prevent neurological disease, although the mechanism(s) of action is as yet unknown.

Interleukin-6 and Glial Cells

Biological Effects of IL-6 on Glial Cells

IL-6 is a pleiotropic cytokine involved in the regulation of inflammatory and immunologic responses. IL-6 has been demonstrated to have multiple effects on the astrocyte, and may function in an autocrine manner as astrogliomas express specific high affinity receptors for IL-6 [182]. IL-6 has a mitogenic effect on astrocytes [67], which may contribute to reactive gliosis. Astrocytes respond to IL-6 by secreting nerve growth factor, which induces neural differentiation [116]. IL-6 itself has been found to induce differentiation of PC 12 cells, as well as increase the number of voltage-dependent Na^+ channels [183]. IL-6 has been shown to inhibit TNF-α production by monocytes [184]. As astrocytes can secrete TNF-α, and TNF-α induces IL-6 production by the astrocyte (see below), this may represent a negative regulatory pathway for controlling TNF-α expression in the CNS.

Another possible role of IL-6 in the CNS may relate to B-cell differentiation. There is ample evidence of B-cell stimulation during various neurological diseases as evidenced by immunoglobulin synthesis within the CNS, and large amounts of immunoglobulin found within CSF [185, 186]. IL-6 has been detected in CSF during viral meningitis and encephalitis [89], and in the CSF of SLE patients having neurological involvement [90]. Increased IL-6 levels have been found in the CNS of mice suffering acute EAE [93], and intracerebroventricular injection of IL-1β in rats induces high circulating levels of IL-6 [187]. Intracerebral production of IL-6 by astrocytes and microglia (see below), the cytokine known to induce the terminal differentiation of B cells into immunoglobulin-secreting plasma cells, may contribute in part to heightened humoral immune responses detected in the CNS during various neurological diseases.

IL-6 Production by Glial Cells

Primary rat and murine astrocytes can secrete IL-6 in response to a variety of stimuli including virus, IL-1, TNF-α, IFN-γ plus IL-1, LPS and calcium ionophore [116, 117, 171]. The human astrocytoma cell line, U373, and glioblastoma line, SK-MG4, express IL-6 mRNA in response to IL-1 [116], while transformed microglia clones also produce IL-6 [126]. As for IL-1 and TNF-α, there are two endogenous CNS sources for IL-6, the astrocyte and microglia. It appears, however, that these two cell types are responsive to different stimuli for IL-6 production as murine microglia do not produce IL-

Table 7. IL-6 and glial cells

Astrocytes
Induces proliferation of astrocytes
Enhances NGF production by primary astrocytes
Astrocytes make IL-6 in response to LPS, IL-1, TNF-α, IFN-γ/IL-1, virus, calcium ionophore
Human glioma cells make IL-6 in response to IL-1
Microglia
Microglia make IL-6 in response to M-CSF, virus

6 in response to IL-1 or TNF-α, whereas murine astrocytes do [116]. Table 7 summarizes the effects of IL-6 on glial cells.

Regulation of IL-6 Gene Expression in Glial Cells

Because of the diverse biological activities of IL-6, and the variety of stimuli by which IL-6 is produced, attention has recently been focused on regulation of IL-6 gene expression. The IL-6 promoter region contains several important transcriptional control element motifs which include: two pairs of glucocorticoid-responsive elements (GRE); an AP-1 binding site, which is present in a number of PMA-inducible promotors; a sequence similar to the human c-fos serum response element (SRE); a core sequence, ACGTCA, of the cyclic AMP-responsive element (CRE), which mediates cAMP inducibility and binds to CREB; and a sequence homologous to the NF-κB binding site [for review, see 103]. A number of studies utilizing the promoter region of the IL-6 gene linked to a reporter gene, as well as deletion mutants of the IL-6 promoter region, have identified several cis-acting elements involved in the inducibility of the IL-6 gene. These include an IL-1 response element (located at –180 to –122) homologous to the c-fos SRE, particularly a 14-bp palindromic sequence (ACATTGCACAATCT), in a human glioblastoma line [188]. Interestingly, we have found that the NF-κB binding site (–73 to –63) is critical for both IL-1 and TNF-α induction of IL-6 in primary rat astrocytes, and that both IL-1 and TNF-α cause nuclear translocation of NF-κB in astrocytes [189]. Thus, it appears that there are differences in the cis-acting DNA elements involved in IL-6 gene regulation in human glioblastoma cells and primary astrocytes. In primary rat astrocytes, although both TNF-α and IL-1β act to induce NF-κB-like nuclear activity and utilize the NF-κB binding site in the IL-6 promoter, they appear to mediate their effects through different second messenger pathways. Preliminary results from our laboratory suggest that IL-1β can induce IL-6 expression through either the cAMP or PKC pathway, while the intracellular signals utilized by TNF-α are presently unknown [Norris and Benveniste, unpubl. observation], demonstrating the complexity of cytokine production in response to different stimuli.

Conclusion

This review highlights the fact that cells of the immune system and CNS can share similar functions: (1) secretion of immunoregulatory cytokines; (2) response to cytokines, and (3) antigen presentation. These properties allow for both physical contact between the two systems, i.e. microglia and/or astrocytes presenting antigen to T cells, as well as communication by soluble factors such as cytokines. There is a complex circuitry of interactions mediated by cytokines, especially in the event of lymphoid/mononuclear cell infiltration into the CNS. The secretion of IFN-γ by infiltrating activated T cells could initially induce astrocytes and microglia to express class I and class II MHC antigens, as well as prime these cells for subsequent cytokine production. The activation of astrocytes and microglia may contribute to either the initiation and/or propagation of intracerebral immune responses. A number of inflammatory mediators such as PGE_2, cytokines such as IFN-α/IFN-β, and endogenous molecules like norepinephrine, can act to ultimately suppress an immune response by inhibiting both class II MHC expression and cytokine production by glial cells. PGE is produced by activated macrophages, microglia and astrocytes, and has been identified immunocytochemically in MS brain [83]. Additionally, astrocytes can produce IFN-α/IFN-β [171, 190], thereby serving as an endogenous source for immunosuppressive molecules.

The induction and ultimate down-regulation of immune responses and cytokine production within the CNS is dependent upon: (1) a dynamic interaction between a variety of peripheral immune cells and CNS cell types; (2) the activational status of these cells; (3) the presence of cytokines with pleiotropic effects (IFN-γ, IL-1, IL-6, TNF-α, etc); (4) the concentration and location of these cytokines in the CNS, and (5) the temporal sequence in which a particular cell responds to the cytokines. The ultimate outcome of immunological and inflammatory events in the CNS will be determined, in part, by an interplay of the above parameters.

The majority of the studies reviewed in this chapter reflect the capacity of cultured primary glial cells or glial cell lines to function in vitro. How these activities correlate to the 'normal' functions of these cells in vivo is difficult to assess. The author would ask, then, that readers consider these studies as potential 'nontraditional' functions of glial cells, especially under pathological conditions. Finally, future studies on the cellular and molecular mechanisms involved with cytokine expression by glial cells, specifically: (1) the nature of the second messenger signals utilized by astrocytes and microglia; (2) delineating specific cis-acting DNA regulatory elements and nuclear proteins involved with MHC and cytokine expression by glial cells, and (3) characterizing cytokine receptors on glial cells, will aid us in better understanding bidirectional communication between the immune system and nervous system.

Acknowledgments

I thank Melissa Mabowitz for her expert secretarial assistance in preparing the manuscript, and thank the members of my laboratory who contributed to some of the studies reported in this chapter, and provided helpful discussion (John R. Bethea, J. Gavin Norris, II Yup Chung, R. Brian Panek, Shaun M. Sparacio, Jong Bum Kwon, Melissa Waldon, Alyssa T. Reddy and Dr. Maria Vidovic). I also thank Drs. Edward Hoover and Steven Dow, Colorado State University, for use of their microglia photomicrographs. These studies were supported by Grant 1954-A-2 from the National Multiple Sclerosis Society (NMSS), and Grant BNS-8708233 from the National Science Foundation, and are currently supported by Grants 2205-A-3 and 2269-A-4 from the NMSS, and NIH Grant Al-27290.

References

1 Peters, A.; Palay, S.L.; Webster, H. de. F.: The Fine Structure of the Nervous System: The Neurons and Supporting Cells, pp. 242–244 (Saunders, Philadelphia 1976).
2 Wong, G.H.W.; Bartlett, P.F.; Clark-Lewis, I.; Battye, F.; Schrader, J.W.: Inducible expression of H-2 and la antigens on brain cells. Nature *310:* 688–691 (1984).
3 Booss, J.; Esiri, M.M.; Tourtelotte, W.W.; Mason, D.Y.: Immunohistological analysis of T lymphocyte subsets in the central nervous system in chronic progressive multiple sclerosis. J. Neurol. Sci. *62:* 219–232 (1983).
4 Prineas, J.W.: Multiple sclerosis. Presence of lymphatic capillaries and lymphoid tissue in the brain and spinal cord. Science *203:* 1123–1125 (1979).
5 Moench, T.R.; Griffin, D.E.: Immunocytochemical identification and quantitation of the mononuclear cells in the cerebrospinal fluid, meninges, and brain during acute viral meningoencephalitis. J. Exp. Med. *159:* 77–88 (1984).

6 Traugott, U.; Reinherz, E.L.; Raine, C.S.: Multiple sclerosis: Distribution of T-cells, T-cell subsets and Ia-positive macrophages in lesions of different ages. J. Neuroimmunol. *4:* 201–221 (1983).
7 Hauser, S.L.; Bhan, A.K.; Gilles, F.H.; Hoban, C. J.; Reinherz, E.L.; Schlossman, S.F.; Weiner, H.L.: Immunohistochemical staining of human brain with monoclonal antibodies that identify lymphocytes, monocytes and the Ia antigen. J. Neuroimmunol *5:* 197–205 (1983).
8 Prineas, J.W.; Wright, R.G.: Macrophages, lymphocytes and plasma cells in the perivascular compartment in chronic multiple sclerosis. Lab. Invest. *38:* 409–421 (1978).
9 Navia, B.A.; Jordan, B.D.; Price, R.W.: The AIDS dementia complex. Ann. Neurol. *19:* 517–524 (1986).
10 Raine, C.S.: Biology of disease: Analysis of autoimmune demyelination: Its impact upon multiple sclerosis. Lab. Invest. *50:* 608–635 (1984).
11 Virchow, R.: Cellular Pathology as Based upon Physiological and Pathological Histology. Edited and translated by F. Chance (Churchill, London 1860).
12 Raff, M.C.: Glial cell diversification in the rat optic nerve. Science *243:* 1450–1455 (1989).
13 Miller, R.H.; Ffrench-Constant, C.; Raff, M.C.: The macroglial cells of the rat optic nerve. Annu. Rev. Neurosci. *12:* 517–534 (1989).
14 Bignami, A.; Eng, L.F.; Dahl, D.; Uyeda, C.T.: Localization of the glial fibrillary acidic protein in astrocytes by immunofluorescence. Brain Res. *43:* 429–435 (1972).
15 Eisenbarth, G.S.; Walsh, F.S.; Nirenberg, M.: Monoclonal antibody to a plasma membrane antigen of neurons. Proc. Natl. Acad. Sci. USA *76:* 4913–4917 (1979).
16 Raff, M.C.; Miller, R.H.; Noble, M.: A glial progenitor cell that develops in vitro into an astrocyte or an oligodeŋdrocyte depending on the culture medium. Nature *303:* 390–396 (1983).
17 Raff, M.C.; Abney, E.R.; Cohen, J.; Lindsay, R.; Noble, M.: Two types of astrocytes in cultures of developing rat white matter. Differences in morphology, surface gangliosides, and growth characteristics. J. Neurosci. *3:* 1289–1300 (1983).
18 Lillien, L.E.; Sentner, M.; Rohrer, H.; Hughes, S.; Raff, M.C.: Type 2 astrocyte development in rat brain cultures is initiated by a CNTF-like protein produced by type-1 astrocytes. Neuron *1:* 485–494 (1988).
19 Norenberg, M.D.; Martinez-Hernandez, A.: Fine structural localization of glutamine synthetase in astrocytes in rat brain. Brain Res. *161:* 303–310 (1979).
20 Sadasivudu, B.; Rao, T.L.; Murthy, C.R.: Acute metabolic effects of ammonia in mouse brain. Neurochem. Res. *2:* 639–645 (1977).
21 Chatterjee, D.; Sarker, P.K.: Ontogeny of glutamine synthetase in rat brain. Int. J. Dev. Neurosci. *2:* 55–60 (1984).
22 Vaccaro, D.E.; Leeman, S.E.; Reif-Lehrer, L.: Glutamine synthetase activity in vivo and in primary cell cultures of rat hypothalamus. J. Neurochem. *33:* 953–957 (1979).
23 Moore, B.W.: A soluble protein characteristic of the nervous system. Biochem. Biophys. Res. Commun. *19:* 739–766 (1965).
24 Ghandour, M.S.; Langley, O.K.; Labourdette, G.; Vencendon, G.; Gombos, G.: Specific and artifactual cellular localization of S-100 protein: An astrocyte marker in rat cerebellum. Dev. Neurosci. *4:* 66–78 (1981).
25 Herschman, H.R.; Levine, L.; de Vellis, J.: Appearance of a brain-specific antigen (S-100 protein) in the developing rat brain. J. Neurochem. *18:* 629–633 (1971).

26 Isobe, T.; Okuyama, T.: The amino acid sequence of S-100 protein (PAP-1-b protein) and its relation to the calcium-binding protein. Eur. J. Biochem. *89:* 379–388 (1978).
27 Johnston, P.V.; Roots, B.I.: Neuron-Glial Relationships. Handbook of Clinical Neurology, pp. 401–421 (North-Holland, Amsterdam 1976).
28 Henn, F.A.; Haljamae, H.; Hamberger, A.: Glial cell function: Active control of extracellular K^+ concentration. Brain Res. *43:* 437–443 (1972).
29 Hansson, E.; Isacsson, H.; Sellstrom, A.: Characteristics of dopamine and GABA transport in primary cultures of astroglial cells. Acta Physiol. Scand. *121:* 333–341 (1984).
30 Schousboe, A.; Hertz, L.; Svenneby, G.: Uptake and metabolism of GABA in astrocytes cultured from dissociated mouse brain hemispheres. Neurochem. Res. *2:* 217–229 (1977).
31 Hertz, L.; Wu, P.H.; Schousboe, A.: Evidence for net uptake of GABA into mouse astrocytes in primary cultures. Its sodium dependence and potassium independence. Neurochem. Res. *3:* 313–324 (1978).
32 Norcnberg, M.D.: The distribution of glutamine synthetase in the rat central nervous system. J. Histochem. Cytochem. *27:* 756–762 (1979).
33 Rakic, P.: Guidance of neurons migrating to the fetal monkey neocortex. Brain Res. *33:* 471–476 (1971).
34 Schmechel, D.E.; Rakic, P.: A Golgi study of radial glial cells in developing monkey telencephalon: Morphogenesis and transformation into astrocytes. Anat. Embryol. *156:* 115–152 (1979).
35 Choi, B.H.; Lapham, L.W.: Interactions of neurons and astrocytes during growth and development of human fetal brain in vitro. Exp. Mol. Pathol. *24:* 110–125 (1976).
36 Janzer, R.C.; Raff, M.C.: Astrocytes induce blood-brain barrier properties in endothelial cells. Nature *325:* 253–257 (1987).
37 Raine, C.S.: Neurons, astrocytes and ependyma; in Davis, Robertson (eds): Textbook of Neuropathology, pp. 468–547 (Williams & Wilkins, Baltimore 1985).
38 Adams, C.W.M.: Multiple sclerosis: pathology, diagnosis and management; in Hallpike, Adams, Tourtellotte (eds): pp. 203–205 (Chapman & Hall, London 1983).
39 Abney, E.R.; Bartlet, P.P.; Raff, M.C.: Astrocytes, ependymal cells and oligodendrocytes develop on schedule in dissociated cell cultures of embryonic rat brain. Dev. Biol. *83:* 301–310 (1981).
40 Norton, W.T.; Cammer, W.: Myelin; in Morell (ed.): Isolation and Characterization of Myelin; 2nd ed., pp. 147–180 (Plenum Press, New York 1984).
41 Lees, M.B.; Brostoff, S.W.: Myelin; in Morell (ed.): Proteins of Myelin; 2nd ed., pp. 197–217 (Plenum Press, New York 1984).
42 Raff, M.C.; Mirsky, R.; Fields, K.L.; Lisak, R.P.; Dorfman, S.H.; Silberberg, D.H.; Gregson, N.A.; Leibowitz, S.; Kennedy, M.C.: Galactocerebroside is a specific cell-surface antigenic marker for oligodendrocytes in culture. Nature *274:* 813–815 (1978).
43 Poduslo, S.E.; Norton, W.T.: Isolation and some chemical properties of oligodendroglia from calf brain. J. Neurochem. *19:* 727–736 (1972).
44 Bloom, F.E.; Battenberg, E.L.; Milner, R.J.; Sutcliffe, J.G.: Immunocytochemical mapping of 1B236. A brain-specific neuronal polypeptide deduced from the sequence of a cloned mRNA. J. Neurosci. *5:* 1781–1802 (1985).

45 Ranscht, B.; Clapshaw, P.A.; Pride, J.; Noble, M.; Seifert, W.: Development of oligodendrocytes and Schwann cells studied with a monoclonal antibody against galactocerebroside. Proc. Natl. Acad. Sci. USA *79:* 2709–2713 (1982).
46 Drummond, G.E.; Iyer, N.T.; Keith, J.: Hydrolysis of ribonucleoside 2′,3′-cyclic phosphates by a diesterase from brain. J. Biol. Chem. *237:* 3535–3539 (1962).
47 Sprinkle, T.J.; Zaruea, M.E.; McKhann, G.M.: Activity of 2′,3′-cyclic-nucleotide 3′-phosphodiesterase in regions of rat brain during development. Quantitative relationship to myelin basic protein. J. Neurochem. *30:* 309–314 (1978).
48 de Vellis, J.; Schjeide, O.A.; Clemente, C.D.: Protein synthesis and enzymatic patterns in the developing brain following head X-irradiation of newborn rats. J. Neurochem. *14:* 499–511 (1967).
49 Breen, G.A.M.; de Vellis, J.: Regulation of glycerol phosphate dehydrogenase by hydrocortisone in rat brain explants. Exp. Cell Res. *91:* 159–169 (1975).
50 Morell, P.; Norton, W.T.: Myelin. Sci. Am. *242:* 88–118 (1980).
51 Bunge, R.P.: Glial cells and the central myelin sheath. Physiol. Rev. *48:* 197–251 (1968).
52 Perry, V.H.; Gordon, S.: Macrophages and microglia in the nervous system. Trends Neurosci. *11:* 273–277 (1988).
53 Hickey, W.F.; Kimura, H.: Perivascular microglial cells of the CNS are bone marrow-derived and present antigen in vivo. Science *239:* 290–292 (1988).
54 Price, R.W.; Brew, B.; Sidtis, J.; Rosenblum, M.; Scheck, A.C.; Cleary, P.: The brain in AIDS: Central nervous system HIV-1 infection and AIDS dementia complex. Science *239:* 586–592 (1988).
55 Ling, E.A.: The origin and nature of microglia; in Federoff, Hertz (eds): Advances in Cellular Neurobiology, vol. 2, pp. 33–82 (Academic Press, New York 1981).
56 Sminia, T.; De Groot, C.J.A.; Dijkstra, C.D.; Koetsier, J.C.; Polman, C.H.: Macrophages in the central nervous system of the rat. Immunobiology *174:* 43–50 (1987).
57 Perry, V.H.; Hume, D.A.; Gordon, S.: Immunohistochemical localization of macrophages and microglia in adult and developing mouse brain. Neuroscience *15:* 313–326 (1985).
58 Giulian, D.; Baker, T.J.: Characterization of ameboid microglia isolated from developing mammalian brain. J. Neurosci. *6:* 2163–2178 (1986).
59 Akiyama, H.; McGeer, P.L.: Brain microglia constitutively express beta-2-integrins. J. Neuroimmunol *30:* 81–93 (1990).
60 Suckling, A.J.; Kirby, J.A.; Rumsby, M.G.: Characterization by acid α-napthyl acetate esterase staining of the spinal cord cellular infiltrate in the acute and relapse phases of chronic relapsing experimental allergic encephalomyelitis. Prog. Brain Res. *59:* 317–322 (1983).
61 Mannoji, H.; Yeger, H.; Becker, L.E.: A specific histochemical marker (lectin *Rincinus communis* agglutinin-I) for normal human microglia, and application to routine histopathology. Acta Neuropathol. *71:* 341–343 (1986).
62 Lassmann, H.; Vass, F.Z.K.; Hickey, W.F.: Microglial cells are a component of the perivascular glia limitans. J. Neurosci. Res. *28:* 236–243 (1991).
63 Adams, C.W.M.: Pathology of multiple sclerosis. Progression of the lesion. Br. Med. J. *33:* 15–20 (1977).
64 Prineas, J.W.: Pathology of the early lesion in multiple sclerosis. Human Pathol. *6:* 531–554 (1975).

65 Giulian, D.; Lachman, L.B.: Interleukin-1 stimulation of astroglial proliferation after brain injury. Science *228:* 497–499 (1985).
66 Giulian, D.; Woodward, J.; Young, D.G.; Krebs, J.F.; Lachman, L.B.: Interleukin-1 injected into mammalian brain stimulates astrogliosis and neovascularization. J. Neurosci. *8:* 2485–2490 (1988).
67 Selmaj, K.W.; Farooq, M.; Norton, W.T.; Raine, C.S.; Brosnan, C.F.: Proliferation of astrocytes in vitro in response to cytokines. A primary role for tumor necrosis factor. J. Immunol. *144:* 129–135 (1990).
68 Raine, C.S.; Scheinberg, C.C.; Waltz, J.M.: Multiple sclerosis: Oligodendrocyte survival and proliferation in an active, established lesion. Lab Invest *45:* 534–546 (1981).
69 Prineas, J.W.; Connell, F.: Remyelination in multiple sclerosis. Ann. Neurol. *5:* 22–31 (1979).
70 Raine, C.S.; Moore, G.R.W.; Hintzen, R.; Traugott, U.: Induction of oligodendrocyte proliferation and remyelination after chronic demyelination. Relevance to multiple sclerosis. Lab. Invest. *59:* 467–476 (1988).
71 Saneto, R.P.; Chiappelli, F.; de Vellis, J.: Interleukin-2 inhibition of oligodendrocyte progenitor cell proliferation depends on expression of the TAC receptor. J Neurosci. Res. *18:* 147–154 (1987).
72 Benveniste, E.N.; Merrill, J.E.: Stimulation of oligodendroglial proliferation and maturation by interleukin-2. Nature *321:* 610–613 (1986).
73 Saneto, R.P.; Altman, A.; Knobler, R.; Johnson, H.M.; de Vellis, J.: Interleukin-2 mediates the inhibition of oligodendrocyte progenitor cell proliferation in vitro. Proc. Natl. Acad. Sci. USA *83:* 9221–9225 (1986).
74 Robbins, D.S.; Shirazi, Y.; Drysdale, B.E.; Leiberman, A.; Shin, H.S.; Shin, M.L.: Production of cytotoxic factor for oligodendrocytes by stimulated astrocytes. J. Immunol. *139:* 2593–2597 (1987).
75 Selmaj, K.W.; Raine, C.S.: Tumor necrosis factor mediates myelin and oligodendrocyte damage in vitro. Ann. Neurol. *23:* 339–346 (1988).
76 Prineas, J.W.; Graham, J.S.: Multiple sclerosis: Capping of surface immunoglobulin-G on macrophages engaged in myelin breakdown. Ann. Neurol. *10:* 149–158 (1981).
77 Epstein, L.B.; Prineas, J.W.; Raine, C.S.: Attachment of myelin to coated pits on macrophages in experimental allergic encephalomyelitis. J. Neurol. Sci. *61:* 341–348 (1983).
78 Moore, G.R.W.; Raine, C.S.: Immunogold localization and analysis of IgG during immune-mediated demyelination. Lab. Invest. *59:* 641–648 (1988).
79 Watkins, B.A.; Dorn, H.H.; Kelly, W.B.; Armstrong, R.C.; Potts, B.J.; Michaels, F.; Kufta, C.V.; Dubois-Dalcq, M.: Specific tropism of HIV-1 for microglial cells in primary human brain cultures. Science *249:* 549–553 (1990).
80 Osborn, L.; Kunkel, S.; Nabel, G.J.: Tumor necrosis factor-α and interleukin-1 stimulate the human immunodeficiency virus enhancer by activation of the nuclear factor κB. Proc. Natl. Acad. Sci. USA *86:* 2336–2340 (1989).
81 Poli, G.; Kinter, A.; Justement, J.S.; Kehrl, J.H.; Bressler, P.; Stanley, S.; Fauci, A.S.: Tumor necrosis factor α functions in an autocrine manner in the induction of human immunodeficiency virus expression. Proc. Natl. Acad. Sci. USA *87:* 782–785 (1990).
82 Wege, H.; Watanabe, R.; ter Meulen, V.: Relapsing subacute demyelinating encephalomyelitis in rats during the course of coronavirus JHM infection. J. Neuroimmunol. *6:* 325–336 (1984).

83 Hofman, F.M.; VonHanwher, R.; Dinarello, C.; Mizel, S.; Hinton, D.; Merrill, J.E.: Immunoregulatory molecules and IL-2 receptors identified in multiple sclerosis brain. J. Immunol. *136:* 3239–3245 (1986).
84 Hofman, F.M.; Hinton, D.R.; Johnson, K.; Merrill, J.E.: Tumor necrosis factor identified in multiple sclerosis brain. J. Exp. Med. *170:* 607–612 (1989).
85 Traugott, U.; Lebon, P.: Interferon-γ and Ia antigen are present on astrocytes in active chronic multiple sclerosis lesions. J. Neurol. Sci. *84:* 257–264 (1988).
86 Selmaj, K.; Raine, C.S.: Cannella, B.; Brosnan, C.F.: Identification of lymphotoxin and tumor necrosis factor in multiple sclerosis lesions. J. Clin. Invest. *87:* 949–954 (1991).
87 Griffin, W.S.T.; Stanley, L.C.; Ling, C.: Brain interleukin-1 and S-100 immunoreactivity are elevated in Down syndrome and Alzheimer disease. Proc. Natl. Acad. Sci. USA *86:* 7611–7615 (1989).
88 Wahl, S.M.; Allen, J.B.; Francis, N.M.: Macrophage- and astrocyte-derived transforming growth factor β as a mediator of central nervous system dysfunction in acquired immune deficiency syndrome. J. Exp. Med. *173:* 981–991 (1991).
89 Frei, K.; Leist, T.P.; Meager, A.; Gallo, P.; Leppert, D.; Zinkernagel, R.M.; Fontana, A.: Production of B cell stimulatory factor-2 and interferon-γ in the central nervous system during viral meningitis and encephalitis. Evaluation in a murine model infection and in patients. J. Exp. Med. *168:* 449–453 (1988).
90 Hirohata, S.; Miyamoto, T.: Elevated levels of interleukin-6 in cerebrospinal fluid from patients with systemic lupus erythematosus and central nervous system involvement. Arthritis Rheum *33:* 644–649 (1990).
91 Maimone, D.; Gregory, S.; Arnason, B.G.W.; Reder, A.T.: Cytokine levels in the cerebrospinal fluid and serum of patients with multiple sclerosis. J. Neuroimmunol. *32:* 67–74 (1991).
92 Symons, J.A.; Bundick, R.V.; Suckling, A.J.; Rumsby, M.G.: Cerebrospinal fluid interleukin-1-like activity during chronic relapsing experimental allergic encephalomyelitis. Clin. Exp. Immunol. *68:* 648–654 (1987).
93 Gijbels, K.; Van Damme, J.; Proost, P.; Put, W.; Carton, H.; Billiau, A.: Interleukin-6 production in the central nervous system during experimental autoimmune encephalomyelitis. Eur. J. Immunol. *20:* 233–235 (1990).
94 Beutler, B.; Cerami, A.: The biology of cachectin/TNF – A primary mediator of the host response. Annu. Rev. Immunol. *7:* 625–655 (1989).
95 Brett, J.; Gerlach, H.; Nawroth, P.; Steinberg, S.; Godman, G.; Stern, D.: Tumor necrosis factor/cachectin increases permeability of endothelial cell monolayers by a mechanism involving regulatory G proteins. J. Exp. Med. *169:* 1977–1991 (1989).
96 Prober, J.S.; Gimbrose, M.A.; Lapierre, L.A.; Mendrick, D.L.; Fiers, W.; Rothlein, R.; Springer, T.A.: Overlapping patterns of activation of human endothelial cells by interleukin-1, tumor necrosis factor, and immune interferon. J. Immunol. *137:* 1893–1896 (1986).
97 Pohlman, T.H.; Stanness, K.A.; Beatty, P.G.; Ochs, H.D.; Harlan, J.M.: An endothelial cell surface factor(s) induced in vitro by lipopolysaccharide, interleukin-1, and tumor necrosis factor-alpha increases neutrophil adherence by a cd18-dependent mechanism. J. Immunol. *136:* 4548–4553 (1986).
98 Smith, C.A.; Davis, T.; Anderson, D.; Solam, L.; Beckmann, M.P.; Jerzy, R.; Dower, S.K.; Cosman, D.; Goodwin, R.G.: A receptor for tumor necrosis factor defines an unusual family of cellular and viral proteins. Science *248:* 1019–1023 (1990).

99 Schall, T.J.; Lewis, M.; Koller, K.J.; Lee, A.; Rice, G.C.; Wong, G.H.; Gatanaga, T.; Granger, G.A.; Lentz, R.; Raab, H.; Kohr, W.J.; Goeddel, D.V.: Molecular cloning and expression of a receptor for human tumor necrosis factor. Cell *61:* 361–370 (1990).

100 Loetscher, H.; Pan, Y.E.; Lahm, H.W.; Gentz, R.; Brockhaus, M.; Tabuchi, H.; Lesslauer, W.: Molecular cloning and expression of the human 55 kd tumor necrosis factor receptor. Cell *61:* 351–359 (1990).

101 de Giovine, F.S.; Duff, G.W.: Interleukin-1: the first interleukin. Immunol. Today *11:* 13–14 (1990).

102 Arai, K.; Lee, F.; Miyajima, A.; Miyatake, S.; Arai, N.; Yokota, T.: Cytokines: Coordinators of immune and inflammatory responses. Annu. Rev. Biochem. *59:* 783–836 (1990).

103 Kishimoto, T.: The biology of interleukin-6. Blood *74:* 1–10 (1989).

104 Le, J.; Vilcek, J.: Biology of disease: Interleukin-6: A multifunctional cytokine regulating immune reactions and the acute phase protein response. Lab. Invest. *61:* 588–602 (1989).

105 van Snick, J.V.: Interleukin-6: an overview. Annu. Rev. Immunol. *8:* 253–278 (1990).

106 Smith, K.A.: Interleukin-2: Inception, impact, and implications. Science *240:* 1169–1176 (1988).

107 Ijzermans, J.N.M.; Marquet, R.L.: Interferon-gamma: A review. Immunobiology *179:* 456–473 (1989).

108 Fontana, A.; Dubs, R.; Merchant, R.; Balsiger, S.; Grob, P.J.: Glia cell stimulating factor (GSF): A new lymphokine. Part 1: Cellular sources and partial purification of murine GSF, role of cytoskeleton and protein synthesis in its production. J. Neuroimmunol. *2:* 55–71 (1982).

109 Fontana, A.; Otz, U.; DeWeck, A.L.; Grob, P.J.: Glia cell stimulating factor (GSF): A new lymphokine. Part 2: Cellular sources and partial purification of human GSF. J. Neuroimmunol. *2:* 73–81 (1982).

110 Merrill, J.E.; Kutsunai, S.; Mohlstrom, C.; Hofman, F.; Groopman, J.; Golde, D.W.: Proliferation of astroglia and oligodendroglia in response to human T-cell derived factors. Science *224:* 1428–1431 (1984).

111 Benveniste, E.N.; Merrill, J.E.; Kaufman, S.E.; Golde, D.W.; Gasson, J.C.: Purification and characterization of human T-lymphocyte derived glial growth promoting factor. Proc. Natl. Acad. Sci. USA *82:* 3930–3934 (1985).

112 Lachman, L.B.; Brown, D.C.; Dinarello, C.A.: Growth promoting effect of recombinant interleukin-1 and tumor necrosis factor for a human astrocytoma cell line. J. Immunol. *138:* 2913–2916 (1987).

113 Frohman, E.M.; Frohman, T.C.; Dustin, M.L.; Vayuvegula, B.; Choi, B.; Gupta, A.; van den Noort, S.; Gupta, S.: The induction of intracellular adhesion molecule 1 (ICAM-1) expression on human fetal astrocytes by interferon-γ, tumor necrosis factor-α, lymphotoxin, and interleukin-1: relevance to intracerebral antigen presentation. J. Neuroimmunol. *23:* 117–124 (1989).

114 Demaine, A.G.: The molecular biology of autoimmune disease. Immunol. Today *10:* 357–361 (1989).

115 Chung, I.Y.; Benveniste, E.N.: Tumor necrosis factor-alpha production by astrocytes: Induction by lipopolysaccharide, interferon-gamma and interleukin-1. J. Immunol. *144:* 2999–3007 (1990).

116 Frei, K.; Malipiero, U.V.; Leist, T.P.; Zinkernagel, R.M.; Schwab, M.E.; Fontana, A.: On the cellular source and function of interleukin-6 produced in the central nervous system in viral diseases. Eur. J. Immunol. *19:* 689–694 (1989).
117 Benveniste, E.N.; Sparacio, S.M.; Norris, J.G.; Grenett, H.E.; Fuller, G.M.: Induction and regulation of interleukin-6 gene expression in rat astrocytes. J. Neuroimmunol. *30:* 201–212 (1990).
118 Tweardy, D.; Mott, P.; Glazer, E.: Monokine modulation of human astroglial cell production of granulocyte colony-stimulating factor and granulocyte-macrophage colony-stimulating factor. I. Effects of IL-1α and IL-1β. J. Immunol. *144:* 2233–2241 (1990).
119 Bethea, J.R.; Chung, I.Y.; Sparacio, S.M.; Gillespie, G.Y.; Benveniste, E.N.: Interleukin-1β induction of tumor necrosis factor-alpha gene expression in human astroglioma cells J. Neuroimmunol. (In press, 1991).
120 Yasukawa, K.; Hirano, T.; Watanabe, Y.; Muratani, K.; Matsuda, T.; Nakai, S.; Kishimoto, T.: Structure and expression of human B cell stimulatory factor-2 (BSF-2/IL-6). EMBO J. *6:* 2939–2945 (1987).
121 Giulian, D.; Baker, T.J.; Shih, L.; Lachman, L.B.: Interleukin-1 of the central nervous system is produced by ameboid microglia. J. Exp. Med. *164:* 594–604 (1986).
122 Fontana, A.; Kristensen, F.; Dubs, R.; Gemsa, D.; Weber, E.: Production of prostaglandin E and an interleukin-1-like factor by cultured astrocytes and C6 glioma cells. J. Immunol. *129:* 2413–2419 (1982).
123 Fontana, A.; Hengartner, H.; de Tribolet, N.; Weber, E.: Glioblastoma cells release interleukin-1 and factors inhibiting interleukin-2-mediated effects. J. Immunol. *132:* 1837–1844 (1984).
124 Roberge, F.G.; Caspi, R.R.; Nusenblatt, R.B.: Glial retinal Müller cells produce IL-1 activity and have a dual effect on autoimmune T helper lymphocytes. J. Immunol. *140:* 2193–2196 (1988).
125 Malipiero, U.V.; Frei, K.; Fontana, A.: Production of hemopoietic colony-stimulating factors by astrocytes. J. Immunol. *144:* 3816–3821 (1990).
126 Righi, M.; Mori, L.; De Libero, G.; Sironi, M.; Biondi, A.; Mantovani, A.; Donini, S.D.; Ricciardi-Castagnoli, P.: Monokine production by microglial cell clones. Eur. J. Immunol. *19:* 1443–1448 (1989).
127 Zamvil, S.S.; Steinman, L.: The T lymphocyte in experimental allergic encephalomyelitis. Annu. Rev. Immunol. *8:* 579–621 (1990).
128 Cross, A.H.; Cannella, B.; Brosnan, C.F.; Raine, C.S.: Homing to central nervous system vasculature by antigen-specific lymphocytes. Lab. Invest. *63:* 162–170 (1990).
129 Gasser, D.L.; Newlin, C.M.; Palm, J.; Gonatas, N.K.: Genetic control of susceptibility to experimental allergic encephalomyelitis in rats. Science *181:* 872–873 (1973).
130 Moore, M.J.; Singer, D.E.; Williams, R.M.: Linkage of severity of experimental allergic encephalomyelitis to the rat major histocompatibility locus. J. Immunol. *124:* 1815–1820 (1980).
131 Williams, R.M.; Moore, M.J.: Linkage of susceptibility to experimental allergic encephalomyelitis to the major histocompatibility locus in the rat. J. Exp. Med. *138:* 775–783 (1973).
132 Mannie, M.D.; Dinarello, C.A.; Paterson, P.Y.: Interleukin-1 and myelin basic protein synergistically augment adoptive transfer activity of lymphocytes mediating

experimental autoimmune encephalomyelitis in Lewis rats. J. Immunol. *138:* 4229–4235 (1987).
133 Jacobs, C.A.; Baker, P.E.; Roux, E.R.; Picha, K.S.; Toivola, B.; Waugh, S.; Kennedy, M.K.: Experimental autoimmune encephalomyelitis is exacerbated by IL-1α and suppressed by soluble IL-1 receptor. J. Immunol. *146:* 2983–2989 (1991).
134 Benveniste, E.N.; Herman, P.K.; Whitaker, J.N.: Myelin basic protein-specific RNA levels in interleukin-2-stimulated oligodendrocytes. J. Neurochem. *49:* 1274–1279 (1987).
135 Benveniste, E.N.; Tozawa, H.; Gasson, J.C.; Quan, S.; Golde, D.W.; Merrill, J.E.: Response of human glioblastoma cells to recombinant interleukin-2. J. Neuroimmunol. *17:* 301–314 (1988).
136 Okamoto, Y.; Minamoto, S.; Shimizu, K.; Mogami, H.; Taniguchi, T.: Interleukin-2 receptor β chain expressed in an oligodendroglioma line binds interleukin-2 and delivers growth signal. Proc. Natl. Acad. Sci. USA *87:* 6584–6588 (1990).
137 Suzumura, A.; Silberberg, D.H.; Lisak, R.P.: The expression of MHC antigens on oligodendrocytes: Induction of polymorphic H-2 expression by lymphokines. J. Neuroimmunol. *11:* 179–190 (1986).
138 Hirayama, M.; Yokochi, T.; Shimokata, K.; Iida, M.; Fujuki, N.: Induction of human leukocyte antigen-A, B, C and -DR on cultured human oligodendrocytes and astrocytes by human γ-interferon. Neurosci. Lett. *72:* 369–374 (1986).
139 Zinkernagel, R.M.; Doherty, P.C.: H-2 compatibility requirement for T cell-mediated lysis of target cells infected with lymphocytic choriomeningitis virus. J. Exp. Med. *141:* 1427–1436 (1975).
140 Skias, D.D.; Kim, D.; Reder, A.T.; Antel, J.P.; Lancki, D.W.; Fitch, F.W.: Susceptibility of astrocytes to class I MHC antigen-specific cytotoxicity. J. Immunol. *138:* 3254–3258 (1987).
141 Fierz, W.; Endler, B.; Reske, K.; Wekerle, H.; Fontana, A.: Astrocytes as antigen presenting cells. I. Induction of Ia antigen expression on astrocytes by T cells via immune interferon and its effect on antigen presentation. J. Immunol. *134:* 3785–3793 (1985).
142 Fontana, A.; Fierz, W.; Wekerle, H.: Astrocytes present myelin basic protein to encephalitogenic T-cell lines. Nature *307:* 273–276 (1984).
143 Pulver, M.; Carrel, S.; Mach, J.P.; de Tribolet, N.: Cultured human fetal astrocytes can be induced by interferon-γ to express HLA-DR. J. Neuroimmunol. *14:* 123–133 (1987).
144 Suzumura, A.; Mezitis, S.G.E.; Gonatas, N.K.; Silberberg, D.H.: MHC antigen expression on bulk isolated macrophage-microglia from newborn mouse brain: Induction of Ia antigen expression by γ-interferon. J. Neuroimmunol. *15:* 263–278 (1987).
145 Massa, P.T.; Dorries, R.; ter Meulen, V.: Viral particles induce Ia antigen expression on astrocytes. Nature *320:* 543–546 (1986).
146 Massa, P.T.; ter Meulen, V.: Analysis of Ia induction on Lewis rat astrocytes in vitro by virus particles and bacterial adjuvants. J. Neuroimmunol. *13:* 259–271 (1987).
147 Pober, J.S.; Gimbrone, M.A.; Cotran, R.S.; Reiss, C.S.; Burakoff, S.J.; Fiers, W.; Rothlein, R.; Springer, T.A.: Ia expression by vascular endothelium is inducible by activated T-cells and by human γ-interferon. J. Exp. Med *157:* 1339–1353 (1983).
148 Male, D.K.; Pryce, G.; Hughes, C.C.W.: Antigen presentation in brain: MHC induction on brain endothelium and astrocytes compared. Immunology *60:* 453–459 (1987).

149 McCarron, R.M.; Kempsi, O.; Spatz, M.; McFarlin, D.E.: Presentation of myelin basic protein by murine cerebral vascular endothelial cells. J. Immunol. *134:* 3100–3103 (1985).

150 Naparstek, Y.; Cohen, I.R.; Fuks, Z.: Vlodavsky, I.: Activated T lymphocytes produce a matrix-degrading heparan sulphate endoglycosidase. Nature *310:* 241–244 (1984).

151 Benveniste, E.N.; Vidovic, M.; Panek, R.B.; Norris, J.G.; Reddy, A.T.; Benos, D.J.: Interferon-γ induced astrocyte class II major histocompatibility complex gene expression is associated with both protein kinase C activation and Na^+ entry. J. Biol. Chem. *266:* 18119–18126 (1991).

152 Traugott, U.; Scheinberg, L.C.; Raine, C.S.: On the presence of Ia-positive endothelial cells and astrocytes in multiple sclerosis lesions and its relevance to antigen presentation. J. Neuroimmunol. *8:* 1–14 (1985).

153 Lee, S.C.; Moore, G.R.W.; Golenwsky, G.; Raine, C.S.: Multiple sclerosis: A role for astroglia in active demyelination suggested by class II MHC expression and ultrastructural study. J Neuropathol. Exp. Neurol. *49:* 122–136 (1990).

154 Rodriguez, M.; Pierce, M.L.; Howie, E.A.: Immune response gene products (Ia antigens) on glial and endothelial cells in virus-induced demyelination. J. Immunol. *138:* 3438–3442 (1987).

155 Sakai, K.; Tabira, T.; Endoh, M.; Steinman, L.: Ia expression in chronic relapsing experimental allergic encephalomyelitis induced by long-term cultured T cell lines in mice. Lab. Invest. *54:* 345–352 (1986).

156 Hickey, W.F.; Osborn, J.P.; Kirby, W.M.: Expression of Ia molecules by astrocytes during acute experimental allergic encephalomyelitis in the Lewis rat. Cell. Immunol. *91:* 528–535 (1985).

157 Matsumoto, Y.; Kawai, K.; Fujiwara, M.: In situ Ia expression on brain cells in the rat: autoimmune encephalomyelitis-resistant strain (BN) and susceptible strain (Lewis) compared. Immunology *66:* 621–627 (1989).

158 Vass, K.; Lassmann, H.; Wekerle, H.; Wisniewski, H.M.: The distribution of Ia antigen in the lesions of rat acute experimental allergic encephalomyelitis. Acta Neuropathol. (Berl.) *70:* 149–160 (1986).

159 Sun, D.; Wekerle, H.: Ia-restricted encephalitogenic T lymphocytes mediating EAE lyse autoantigen-presenting astrocytes. Nature *320:* 70–72 (1986).

160 Sasaki, A.; Levison, S.W.; Ting, J.P.Y.: Differential suppression of interferon-γ-induced Ia antigen expression on cultured rat astroglia and microglia by second messengers. J. Neuroimmunol. *29:* 213–222 (1990).

161 Frohman, E.M.; Vayuvegula, B.; Gupta, S.; van den Noort, S.: Norepinephrine inhibits γ-interferon-induced major histocompatibility class II (Ia) antigen expression on cultured astrocytes via β_2-adrenergic signal transduction mechanisms. Proc. Natl. Acad. Sci. USA *85:* 1292–1296 (1988).

162 Massa, P.T.; ter Meulen, V.; Fontana, A.: Hypersensitivity of Ia antigen on astrocytes correlates with strain-specific susceptibility to experimental autoimmune encephalomyelitis. Proc. Nat. Acad. Sci. USA *84:* 4219–4223 (1987).

163 Massa, P.T.; Brinkmann, R.; ter Meulen, V.: Inducibility of Ia antigen on astrocyte by murine coronavirus JHM is rat strain-dependent. J. Exp. Med. *166:* 259–264 (1987).

164 Benveniste, E.N.; Sparacio, S.M.; Bethea, J.R.: Tumor necrosis factor-α enhances interferon-γ mediated class II antigen expression on astrocytes. J. Neuroimmunol. *25:* 209–219 (1989).

165 Bethea, J.R.; Gillespie, G.Y.; Chung, I.Y.; Benveniste, E.N.: Tumor necrosis factor production and receptor expression by a human astroglioma cell line. J. Neuroimmunol. *30:* 1–13 (1990).
166 Shin, M.L.; Koski, C.L.: The complement system in demyelination; in Martenson (ed.): Myelin: A Treatise (Telford Press, New Jersey, in press, 1991).
167 Levi-Strauss, M.; Mallat, M.: Primary cultures of murine astrocytes produce C3 and factor B, two components of the alternative pathway of complement activation. J. Immunol. *139:* 2361–2366 (1987).
168 Barnum, S.R.; Jones, J.L.; Benveniste, E.N.: Interferon-gamma regulation of C3 gene expression in human astroglioma cells. J. Neuroimmunol. (in press, 1991).
169 Panitch, H.S.; Hirsch, R.L.; Schindler, J.; Johnson, K.P.: Treatment of multiple sclerosis with gamma-interferon: Exacerbations associated with activation of the immune system. Neurology *37:* 1097–1102 (1987).
170 Billiau, A.; Heremans, H.; Vandekerckhove, F.; Dijkmans, R.; Sobis, H.; Meulepas, E.; Carton, H.: Enhancement of experimental allergic encephalomyelitis in mice by antibodies against IFN-γ. J. Immunol. *140:* 1506–1510 (1988).
171 Lieberman, A.P.; Pitha, P.M.; Shin, H.S.; Shin, M.L.: Production of tumor necrosis factor and other cytokines by astrocytes stimulated with lipopolysaccharide or a neurotropic virus. Proc. Natl. Acad. Sci. USA *86:* 6348–6352 (1989).
172 Lavi, E.; Suzumura, A.; Murasko, D.M.; Murray, E.M.; Silberberg, D.H.; Weiss, S.R.: Tumor necrosis factor induces expression of MHC class I antigens on mouse astrocytes. J. Neuroimmunol. *18:* 245–253 (1988).
173 Massa, P.T.; Schmipl, A.; Wecker, E.; ter Meulen, V.: Tumor necrosis factor amplifies virus-mediated Ia induction on astrocytes. Proc. Natl. Acad. Sci. USA *84:* 7242–7245 (1987).
174 Vidovic, M.; Sparacio, S.M.; Elovitz, M.; Benveniste, E.N.: Induction and regulation of class II MHC mRNA expression in astrocytes by IFN-γ and TNF-α. J. Neuroimmunol. *30:* 189–200 (1990).
175 Jirik, F.R.; Podor, T.J.; Hirano, T.; Kishimoto, T.; Loskutoff, D.J.; Carson, D.A.; Lotz, M.: Bacterial lipopolysaccharide and inflammatory mediators augment IL-6 secretion by human endothelial cells. J. Immunol. *142:* 144–147 (1989).
176 Frei, K.; Siepl, C.; Groscurth, P.; Bodmer, S.; Schwerdel, C.; Fontana, A.: Antigen presentation and tumor cytotoxicity by interferon-γ-treated microglial cells. Eur. J. Immunol. *17:* 1271–1278 (1987).
177 Sinha, A.A.; Lopez, M.T.; McDevitt, H.O.: Autoimmune diseases: The failure of self-tolerance. Science *248:* 1380–1388 (1990).
178 Müller, U.; Jongeneel, C.V.; Nedospasov, S.A.; Lindahl, K.F.; Steinmetz, M.: Tumour necrosis factor and lymphotoxin genes map close to H-2D in the mouse major histocompatibility complex. Nature *325:* 265–267 (1987).
179 Chung, I.Y.; Norris, J.G.; Benveniste, E.N.: Differential TNF-α expression by astrocytes from experimental allergic encephalomyelitis susceptible and resistant rat strains. J. Exp. Med. *173:* 801–811 (1991).
180 Powell, M.B.; Mitchell, D.; Lederman, J.; Buckmeier, J.; Zamvil, S.S.; Graham, M.; Ruddle, NH.; Steinman, L.: Lymphotoxin and tumor necrosis factor-alpha production by myelin basic protein-specific T cell clones correlates with encephalitogenicity. Int. Immunol. *2:* 539–544 (1990).
181 Ruddle, N.H.; Bergman, C.M.; McGrath, K.M.; Lingenheld, E.G.; Grunnet, M.L.; Padula, S.J.; Clark, R.B.: An antibody to lymphotoxin and tumor necrosis factor

prevents transfer of experimental allergic encephalomyelitis. J. Exp. Med. *172:* 1193–1200 (1990).

182 Taga, T.; Kawanishi, Y.; Hardy, R.R.; Hirano, T.; Kishimoto, T.: Receptors for B cell stimulatory factor 2. Quantitation, specificity, distribution, and regulation of their expression. J. Exp. Med. *166:* 967–981 (1987).

183 Satoh, T.; Nakamura, S.; Taga, T.; Matsuda, T.; Hirano, T.; Kishimoto, T.; Kaziro, Y.: Induction of neuronal differentiation in PC12 cells by B-cell stimulatory factor 2/interleukin-6. Mol. Cell. Biol. *8:* 3546–3549 (1988).

184 Aderka, D.; Le, J.; Vilcek, J.: IL-6 inhibits lipopolysaccharide-induced tumor necrosis factor production in cultured human monocytes, U937 cells, and in mice. J. Immunol. *143:* 3517–3523 (1989).

185 Tourtellotte, W.W.; Ma, I.B.: Multiple sclerosis: The blood-brain barrier and the measurement of de novo central nervous system IgG synthesis. Neurology *28:* 76–83 (1978).

186 Resnick, L.; diMarzo-Veronese, F.; Schupbach, J.; Tourtellotte, W.W.; Ho, D.D.; Müller, F.; Shapshak, P.; Vogt, M.; Groopman, J.E.; Markham, P.D.; Gallo, R.C.: Intra-blood-brain barrier synthesis of HTLV-III-specific IgG in patients with neurologic symptoms associated with AIDS or AIDS-related complex. N. Engl. J. Med. *313:* 1498–1504 (1985).

187 de Simoni, M.G.; Sironi, M.; de Luigi, A.; Manfridi, A.; Mantovani, A.; Ghezzi, P.: Intracerebroventricular injection of interleukin-1 induces high circulating levels of interleukin-6. J. Exp. Med. *171:* 1773–1778 (1990).

188 Isshiki, H.; Akira, S.; Tanabe, O.; Nakajima, T.; Shimamoto, T.; Hirano, T.; Kishimoto, T.: Constitutive and interleukin (IL-1)-inducible factors interact with the IL-1-responsive element in the IL-6 gene. Mol. Cell. Biol. *10:* 2757–2764 (1990).

189 Sparacio, S.M.; Zhang, Y.; Vilcek, J.; Benveniste, E.N.: Regulation of interleukin-6 gene expression in astrocytes involves activation of an NF-κB-like nuclear protein (submitted, 1991).

190 Tedeschi, B.; Barrett, J.N.; Keane, R.W.: Astrocytes produce interferon that enhances the expression of H-2 antigens on a subpopulation of brain cells. J. Cell Biol. *102:* 2244–2253 (1986).

Etty N. Benveniste, PhD, Departments of Cell Biology and Neurology,
University of Alabama at Birmingham, UAB Station, Birmingham, AL 35294 (USA)

Blalock JE (ed): Neuroimmunoendocrinology, 2nd rev ed.
Chem Immunol. Basel, Karger, 1992, vol 52, pp 154–169

Hormonal Activities of Cytokines

Eric M. Smith

Departments of Psychiatry and Microbiology,
University of Texas Medical Branch, Galveston, Tex., USA

Introduction

The lymphokine and cytokine field has been steadily expanding since the initial version of this chapter was written in 1987 [1]. Since then a number of new lymphokines and monokines have been identified, so that in the nomenclature scheme, the recently described hemopoietic growth factor is interleukin-11 (IL-11) [2]. In light of this systemic naming and in the interest of simplicity, all these factors: lymphokines, monokines, interferons (IFNs), etc., will be referred to by the broader term cytokines. For a recent listing of the major cytokines produced by the immune system and their activities, see Balkwill and Burke [3]. These have all been characterized functionally and most have been cloned and sequenced, at least at the cDNA level. A hallmark of the soluble mediators in the immune system is to be multifunctional. A trend that has become apparent and is relevant to a chapter such as this is that, with time, multiple functions have been identified for all the cytokines [3]. Thus it should come as no surprise that these pluripotent molecules also have many systemic effects.

In such a rapidly expanding field as cytokines, it is too soon for much transfer of the newer molecules into the study of neuroendocrinimmunology. Most of the recent studies have continued with IL-1, 2, 3 and the IFNs. An anatomical fact that should be kept in mind when assessing the role of cytokines on neuroendocrine tissues is that most, if not all, endocrine glands contain a high number of lymphoid cells, up to 20% of the total cell number in the case of the adrenal gland [4]. This, coupled with data showing that neuroendocrine tissues synthesize various cytokines, makes it very likely that endocrine tissues are routinely exposed to high concentrations of cytokines.

Interleukin-1

IL-1, also known as the endogenous pyrogen, has the most systemic and neuroendocrine type effects of the cytokines, with the possible exception of the IFNs [5, 6]. IL-1 is produced by all types of macrophages such as alveolar and peritoneal macrophages plus Kupffer cells. Now it is recognized that virtually all nucleated cell types – keratinocytes, epithelial cells, astrocytes, microglial cells, fibroblasts, and many transformed cells – produce IL-1-like factors [5, 6]. Two species of IL-1 exist: -α and β. They appear to act similarly and in the brain bind equally to the same receptor [7, 8] and therefore, except for the rare cases when they differ, the two will be referred to collectively. The pyrogenic and somnogenic effects of IL-1 as well as activation of the hypothalamic-pituitary-adrenal (HPA) axis vividly illustrate the broad range of activities associated with this cytokine. The ability of microglial cells and brain macrophages to produce IL-1, coupled with its presence in neurons, proves that this cytokine is not restricted by blood-brain barriers (BBB) [8–10].

Neuroendocrine Action

Previously, there was a great controversy concerning whether IL-1 acted at the pituitary or hypothalamus sites [1, 11–18]. There are many conflicting reports, but most of the studies have been replicated and with the identification of IL-1 receptors on pituitary cells [8, 10], it must be surmised that IL-1 is capable of acting at both levels. We originally reported that IL-1 and hepatocyte-stimulating factor (now known to be IL-6) stimulate the mouse pituitary tumor cell line, AtT-20, to produce ACTH [11]. IL-1 was equipotent with corticotropin-releasing hormone (CRF). Bernton et al. [14] found that primary isolated pituitary cell cultures also respond to IL-1 by releasing ACTH, growth hormone (GH), luteinizing hormone (LH) and thyroid-stimulating hormone (TSH), while prolactin (PRL) secretion was inhibited. The findings of IL-1 stimulation of AtT-20 cells [11, 14] and primary pituitary cells to produce ACTH has been reproduced by several laboratories [16, 18–22] with some interesting variations that identify some of the factors that may be confusing the issue. Certainly the choice of cells is one factor. AtT-20 cells are a tumor cell line and may obviously differ in their response from primary pituitary cells. Even this cell line varies, some strains grow in monolayers, whereas the original ATCC strain we used grows as a suspension culture. These phenomena are not limited to IL-1. Fukata et al. [17] found their AtT-20 cells to be insensitive to CRF, unless pretreated with IL-1. Also, these cells required 4 h or more for stimulation of ACTH release. Similarly, Fagarasan et al. [21] found an 18–24 h latent period necessary for their AtT-20 cells to respond to IL-1α and β. However, IL-1 would enhance CRF induction of ACTH secretion by incubation for 12 h. This is in contrast to our studies in which significant ACTH was released in 2 h [11]. Perhaps with serial passage there are changes in surface receptor expression. Thus, without a serious effort to optimize the system, a false-negative result could easily be obtained even when the procedures are seemingly equal.

Presumably the same factors, in conjunction with the difficulties of primary cell culture, have contributed to the mixed results obtained with fresh pituitary cells. Cell culture and perfusion studies from Holiday and Bernton's [14] group conferred on multiple hormone release, but the latter did not alter PRL release. Others have found IL-1 had no effect or else large doses were required [12, 13, 15, 18]. In one of the few studies using human cultured pituitary cells, IL-1β stimulated ACTH release, but little or no GH, PRL, TSH, LH or FSH [19]. The pituitary cells were from patients with Cushing's disease however, and therefore there might have been an altered response. The cellular oncogenes c-fos and c-jun are activated early in AtT-20 cells by IL-1 [22].

Using superfused anterior pituitary lobes, Boyle et al. [23] found IL-1 prevented the loss of corticotropic responsiveness to β-adrenergic stimulation. The authors postulate that IL-1 in vivo may function to maintain β-adrenergic responsiveness. This loss of responsiveness was evident after 4 h. A rapid change in response during culture could also explain negative results in other studies. The presence of IL-1 receptors on pituitary cells [8, 10] and the fact that ACTH release can be prostaglandin E_2 (PGE_2) [20] and catecholamine [24] independent substantiates that IL-1 may have a direct if not major effect on the pituitary in vivo.

Injection of IL-1 into intact animals activates the HPA axis to induce steroidogenesis [15, 25, 26]. In the same issue of *Science* as the Bernton et al. [14] paper, Sapolsky et al. [12] showed in an elegant experiment that antibody to CRF blocked corticosteroid induction when IL-1 was injected intravenously. Berkenbosch et al. [13] reported that intraperitoneal injection of IL-1 activated rat CRF-producing neurons. If the animals are deafferentiated, the IL-1 induction of corticosterone is inhibited [26]. Hypothalamic slices stimulated in vitro with IL-1 also release CRF [27]. Thus it may be that in the intact animal, the major or initial effect of IL-1 is mediated through the hypothalamus.

Another possible site of action of IL-1 in stimulating adrenal steroidogenesis is the adrenal cortical cells themselves. We were unable to induce cultured mouse Y-1 adrenal cells to release glucocorticoid hormones in response to IL-1 or IL-6 [11]. Bartfai et al. [28] have recently reported that IL-1 is present in significant quantities in adrenal medulla noradrenergic chromaffin cells. These cells are located at the rim of the gland and in proximity to the adrenal cortex and therefore might affect steroidogenesis. Winter et al. [29] found that IL-1α and β would induce cultured bovine adrenal cells to secrete cortisol. However, indomethacin inhibited this secretion so an indirect mechanism through prostaglandins was postulated. Dyck et al. [30] have reported IL-1-induced glucocorticoid secretion is significantly enhanced through Pavlovian conditioning. Thus, there appear to be many factors and sites mediating IL-1 stimulation of glucocorticoids.

Other neuroendocrine-like actions are continually being described. Bernton et al. [14] showed that TSH, GH, PRL, and LH release by pituitary cells was affected by IL-1. With some exceptions, in particular hormones [18] or in pituitary cells from patients with Cushing's disease [19], IL-1 has been shown to be pluripotent. Like other cytokines it has been suggested to play a role in pregnancy. It will inhibit luteinization of porcine granulosa cells [31, 32], stimulate chorionic gonadotropin secretion by trophoblast cells [33] and induce postimplantation pregnancy failure in mice after intracerebroventricular administration [34]. Finally, inhibition of pancreatic islet function [35] suggests IL-1 may also play a role in diabetes.

Central Nervous System (CNS) Action

The presence of IL-1 mRNA [9], protein [10, 36, 37] and receptors [7, 38, 39] in the brain suggest that it may not be necessary for IL-1 to cross the BBB to have CNS effects. Its presence and the evolutionary conservation of this relationship [37] implies that it might serve certain endogenous CNS functions.

Multiple CNS effects of IL-1 have been reported. Fever induction [40] and enhancement of slow wave sleep [41] are two examples that are inducible through systemic or central administration of IL-1. Some of the early CNS studies in the field performed by Besedovsky et al. [42, 43] looking at alterations in neuronal firing rates and neurotransmitter turnover rates during immune responses can probably be attributed, at least in part, to IL1. Later studies showed that systemic injection of IL-1 did induce the release of glucocorticoid hormones [25] and that CRH-producing neurons were activated [13].

Reports of specific IL-1 receptors in the brain support these functional studies and also suggest new CNS regions that may be affected by this cytokine [7, 38, 39]. IL-1 receptors are distributed widely over the entire brain, but certain regions express higher concentrations of the receptor. Regions rich in IL-1 receptors include the hippocampus, olfactory bulb, cerebellar cortex, choroid plexus, ventromedial hypothalamus, and anterior preoptic area, among others [7]. Interestingly, the anterior hypothalamus has a low density of IL-1 binding. This region, associated with the posterior preoptic area, is the region from which the most rapid and intense febrile response to locally injected IL-1 is induced and might be expected to be rich in binding sites [40]. Since the choroid plexus is densely populated with IL-1 receptors, this argues that systemic IL-1 may act here and not cross the BBB, as labeling studies suggest [40]. Blatteis [40] has suggested that entry might be gained through the circumventricular organs, which lack a BBB. He further points out, '... that the organum vasculosum laminae (OVLT) is located in the center of the preoptic-anterior hypothalamus and the medullary IL-1-sensitive site lies close to the area postrema.' Experimental support for this idea is that ablation of the OVLT attenuates the fever caused by injection of endotoxin or IL-1, IFN, and tumor necrosis factor (TNF) [40]. Endotoxin also alters IL-1 receptor expression [38]. The receptor down-regulation is probably due to IL-1 induced by the endotoxin as there was no decrease in receptors in mice that do not produce IL-1. The density of IL-1 receptors was significantly reduced in the dentate gyrus and the choroid plexus, without changing the K_d, after intravenous injection of endotoxin. However, pituitary IL-1 receptor levels remained constant.

A possible indirect mechanism for IL-1 action is through prostaglandins. IL-1 will induce lymphoid, neural and fibroblast cells to produce E series prostaglandins (PGE) [24, 44, 45]. PGE induces many of the same effects as IL-1 such as ACTH production by pituitary cells [46] and fever [40]. Several studies have used PGE inhibitors to suggest that IL-1 action is independent, so either the mechanism is resistant to these inhibitors or another mechanism is at work. Alternatively, central amines have also been postulated to play an indirect role [6].

No unifying role of IL-1 in nervous and immune system interactions has been identified. The number and variety of hormonal and neural actions of IL-1 suggests there will be no single role. The apparent conservation of IL-1 and especially its neural location suggests that this relationship is of fundamental importance [37, 47].

Interleukin-2

IL-2 is the prototypic lymphotrophic hormone which is produced primarily by T lymphocytes of the T_H1 subtype and originally identified as T cell growth factor [48]. It is the product of a single gene and induced through stimulation by IL-1 or other T cell mitogenic agents. In the immune system IL-2 is associated with production of lymphokines (IFN, CSF, IL-3, IL-4 and lymphotoxins), growth of cytotoxic lymphocytes, and the induction of lymphokine-activated killing (LAK cells) and natural killer activities [3]. This lymphokine has had a major impact on the study of immunology, introducing endocrine, pharmacological, and intracellular messenger concepts to the field. Consequently, much is known concerning its immune functions, but ironically little is known of its endocrine effects.

Early indications that IL-2 had hormonal activity were implied from the same experiments cited above for IL-1 by Besedovsky et al. [42, 43, 49]. The supernatant fluids from mitogen-treated lymphocyte cultures probably also contained IL-2 as well as IL-1, and the glucocorticoid induction caused by these fluids might have been attributed to either. Clinical trials to induce LAK antitumor activity [50–52] showed purified IL-2 had inherent endocrine-like systemic effects. Large doses of IL-2 had many toxic side effects, comparable to those induced by IFN, such as fever, fatigue, hypotension, myocardial arrhythmias and neuropsychiatric effects [53–55].

In vitro studies have supported these clinical findings and shown direct effects on pituitary hormone production. Stimulation of pituitary cells and pituitary tumor cells indicates that IL-2 enhances POMC gene expression [16]. Using hemipituitaries, Karanth and McCann [56] expanded these in vitro findings to other neuropeptides. There was a differential effect on release of the different pituitary hormones. ACTH, PRL and TSH release was rapid and confirmed the previous study by Brown et al. [16]. Conversely, IL-2 inhibited the basal release of LH, FSH and GH. The authors pointed out that except for TSH, the pattern of hormone release resembled that seen in vivo with stress. Since IL-1 induces T cells to produce IL-2, these responses might represent a possible feedback mechanism in response to IL-1 effects. It will be important to determine if IL-2 also acts at the hypothalamic level.

In this regard the bulk of the data on IL-2 in this field is on its location in the brain and its neural properties. The localization of IL-2 and its receptors in the brain, especially in regard to neural infections and autoimmune disease, is discussed in detail elsewhere in this volume.

Other Interleukins and IL-6

Even after 4 years, little is known about the hormonal activities of the other interleukins IL-3 to -11. This is partly due to the recent identification and limited availability of these factors. IL-3 is produced by astrocytes and may have a role in the CNS as discussed elsewhere in this volume.

IL-6, a monokine, has many similarities to IL-1, both structurally and functionally [for review, see 57, 58]. IL-6 was identified independently by several laboratories and has been variously termed B cell stimulatory factor, hepatocyte-stimulating factor and IFN-β_2, depending upon the system in which it was identified. In keeping with its many functions, there appear to be multiple pathways regulating IL-6 production. One of these involves induction of IL-6 by IL-1 and IL-2 [57]. Like IL-1, many cell types produce IL-6; most macrophage-like cells (i.e. microglial, Kuppfer cells), B cells, T cells, fibroblasts and epithelial cells also produce IL-6. Thus, in any interaction in which IL-1 is involved, IL-6 must also be considered.

In an acute phase response, such as endotoxic shock, IL-6, TNF and IL-1 are involved in a systemic activation which results in fever plus activation of the HPA axis. As reviewed previously [1], we found IL-6 (thought to be hepatocyte-stimulating factor in our study) to be a more potent secretagogue for ACTH by AtT-20 pituitary tumor cells than CRF [11]. Since glucocorticoids inhibit its and IL-1's production [59], it appeared that these monokines might be part of a bidirectional regulatory loop between the HPA axis and the immune system during an acute phase response. Studies in the last few years have confirmed these results and suggested specific mechanisms for such a circuit.

Intravenous injection of IL-6 results in a dose-dependent increase in plasma ACTH levels [60]. As seen with IL-1 [12], antibody to CRF blocked the rise in ACTH induced by IL-6 [60]. Even though IL-6 stimulates ACTH, GH and PRL release by pituitary cells in vitro [17, 58], it seems circulating IL-6 acts indirectly. The lymphoid system is not the only source of IL-6 and it happens that the nervous and neuroendocrine systems synthesize IL-6 [58, 61–63]. De Simoni et al. [64] injected IL-1 intracerebroventricularly and found increased levels of IL-6, probably from the brain. Spangelo et al. [63] then found IL-6 to be produced by cultured pituitaries and the production enhanced by LPS or phorbol myristate. IL1α and IL-1β also significantly increased pituitary IL-6 production in the same manner as in nonlymphoid cells [58]. Pituitary IL-6 production was inhibited by dexamethasone [58] showing that positive and negative regulation of IL-6 production is similar in the neuroendocrine and immune system. IL-6 production may vary with development since young rat pituitaries produce much less IL-6 than do older pituitaries.

Another likely source of IL-6 that could stimulate pituitary cells is the hypothalamus [62]. Yamaguchi et al. [65] report that primary cultures of hypothalamic neurons release gonadotropin-releasing hormone and IL-6 when stimulated by estrogen. Spangelo et al. [62] report that endotoxin induces the release of IL-6 from medial basal hypothalami in vitro. Reports of astrocytes and microglial cells producing IL-6 verify the neural association of IL-6. We recently found that human IL-6 will potentiate IL-1 activation of invertebrate immunocytes [Hughes et al., manuscript submitted]. Thus, IL-6 functioning appears to have been conserved, like IL-1, and that these cytokines must serve some fundamental purpose.

Tumor Necrosis Factor

TNF shares many activities and interactions with IL-1 and IL-6 [66]. Labeled as cachectin, TNF is produced during endotoxic shock and is thought to mediate many of the systemic effects seen in this syndrome. The wasting or cachexia is a catabolic state that can be induced in vitro in adipocytes [66]. TNF and IFN-γ have been implicated in vitro as cytotoxic to pancreatic islet β-cells [67]. However, in diabetic rodents, TNF acts in an opposite fashion and actually inhibits type 1 diabetes [68]. Further indications of the importance of cytokines in diabetes are recent findings that islet cells from diabetic rats express cytokine mRNA [69]. Forty percent of islet cells in this model express IL-1 and TNF mRNA and 4% express IL-6 mRNA. At the neuroendocrine level, TNF inhibits GH secretion from cultured pituitary cells, so it is likely that TNF's metabolic alterations occur at many levels [70].

TNF is synthesized by astrocytes [71] and it seems likely that it may also be produced by pituitary cells, like IL-1 and IL-6 [63]. TNF acts at the hypothalamic-pituitary level by stimulating adrenal and inhibiting thyroid functions [72]. TNF will also act at the level of the effector gland to inhibit thyroid function. Stimulation of α-adrenergic receptors augments the production of macrophage-derived TNF [73]. TNF may also be involved in pregnancy. TNF mRNA and protein have been found in rat uterine and placental cells [74]. This variety of findings suggests that TNF has hormonal-like effects and its regulation may respond to hormones, thus providing the basis of another regulatory loop.

Interferons

Identified in 1957 as an antiviral protein, IFN is now known to consist of at least three major types (α, β, γ) and numerous subtypes. Although the IFNs have antiviral properties, current research seems to be stressing their immunoregulatory roles. Numerous hormonal activities have been described for the IFNs [for review, see 1], which seem to explain many of the toxic side effects seen when large doses of IFN were administered in clinical trials [53–55]. IFN-α has been found to activate several endocrine systems, apparently through common intracellular pathways [1, 75–78]. Since most cells possess IFN receptors, the spectrum of hormonal actions is likely to be quite large.

IFN-α has been shown to affect tumor cell growth, myocardial cell beat frequency, corticosteroid production, cell morphology, opiate receptor binding, and melanogenesis, among other effects [1, 75–78]. In vivo, IFN demonstrates related effects. Dependent upon the dose, lethargy, increased slow wave sleep, analgesia, increased cortisol release, fever, and mental impairment have all been reported [53–55]. The neurological effects may be due to IFN-α's ability to bind to opiate receptors or mediated through IFN receptors [79–84].

As the clinical use of IFN has increased, the general clinical trials and more global basic studies have become narrower in scope and fewer studies have addressed IFN's hormonal activity. Action in the HPA axis is a general feature of all the IFN types. Patients with chronic active hepatitis treated with recombinant IFN-α had significant rises in GH, ACTH, and cortisol [85]. Billiau and Denef [86] found that in contrast to IFN-α, IFN-γ inhibited CRH-induced ACTH secretion by cultured pituitary cells. In their system, TNF-α and IL-6 also inhibited ACTH secretion which contrasts with previous studies in which these cytokines were stimulatory in the absence of CRH. Corticosteroids tend to suppress immune functions and might mediate increased susceptibility to infection during times of stress [87]. IFN-γ has a positive effect on neutrophil function in dexamethasone-treated cattle and thus might help combat 'shipping stress' [88]. Injection of IFN-α into the third cerebral ventricle induced ACTH release and lowered GH and TSH levels [89]. IFN-α also stimulated ACTH release from hemipituitaries in vitro.

The neural activity of IFN has continued to be of interest. Previous studies stressed the opiate-like activity of IFN-α [1]. With minor differences, this has held up and been expanded to more specifically identify the various sites of activity. IFN-α and IL-1β inhibit the increased activity of glucose-responsive neurons in the ventromedial hypothalamus [90].

Conflicting results have been reported on the effect of estrogen and IFN antitumor activity [91, 92]. One group finds no relationship between estrogen and IFNs using cultured breast cells [91], while a study measuring oligoadenylate synthetase activity in chronic lymphocytic leukemia cells finds a synergistic effect [92].

Thus, with little new information and major differences in the published studies, it is clear that much needs to be done to fully understand the hormonal activities of IFNs.

Conclusion

In updating this chapter, several aspects of the topic became clear. Not too surprising is that with time more activities, immunological and hormonal, have been found for each cytokine. However, the monokines seem to have a disproportionately large number of extraimmune activities. This is certainly logical if required to support the diverse actions and locations for monocyte/macrophage lineage cells. A prime example of this may be microglial cells, which are a major source of these cytokines in the nervous system, express a number of immunological features, and are probably the cells of the CNS infected by the AIDS virus.

Another interesting trend is the acceptance of this branch of neuroendocrinimmunology. A number of established, conventional immunologists and endocrinologists are publishing on hormonal and neural activities of cytokines. With them have come the detailed studies, which are really convincing that there are many signal molecules and receptors common to the immune and neuroendocrine systems.

Another aspect illustrated throughout this chapter has been multiplicity, both in molecules and in function. Experiments using a single cytokine, administered as a bolus, to treat a single cultured cell line can only provide a limited perspective of what has to be multifactorial in vivo. The hormonal environment of the circulatory system, the fact that cytokines may induce other cytokines and hormones, circadian rhythms, and common intracellular pathways, are all mediating factors.

Normal functioning of these cytokines within the immune system is often dependent upon multiple signals. Considering the multiple cytokines and neuropeptides that have effects individually, this is becoming an important issue for understanding neuroendocrine interactions within the immune system [see Roszman, this volume]. Recent reports illustrate the significance of understanding these types of complex interactions; some involve multiple signals which act coordinately on the same cell while others act sequentially and on different cells. For instance, glucocorticosteroid-mediated inhibition of T cell proliferation can be abrogated by a synergistic combination of IL-1, IL-6, and IFN-γ, but not by any individually [93]. As an example of the second system, IL-1 is an intermediate in CRH induction of ACTH by lymphocytes [94]. Further complicating the system is that even the endocrine target cells may produce these factors. For instance, both IL-1 [28] and IL-6 [95] are produced in the adrenal.

Thus, until we can better understand the integrated functioning of these molecules, it is impossible to draw any conclusion as to a particular cytokine's true hormonal function. That these interactions are significant is supported by the findings of similar neuropeptide/cytokine relationships in invertebrates [47]. Certainly, conservation of these molecules and interactions over millions of years of evolution implies that these are fundamental mechanisms.

Acknowledgments

I would like to express my gratitude to Judy Van Over, Nita Brannon and Anne Millard for their invaluable assistance in preparation of the manuscript. The studies from my laboratory referred to in this article were supported in part by the Office of Naval Research (N00014-89-1095) and the National Institutes of Health (DK41034).

References

1 Smith EM: Hormonal activities of lymphokines, monokines, and other cytokines. Prog Allergy 1988;43:121–139.
2 Durum SK, Quinn DG, Muegge K: New cytokines and receptors make their debut in San Antonio. Immunol Today 1991;12:54–57.
3 Balkwill FR, Burke F: The cytokine network. Immunol Today 1989;10:299–304.
4 Hume DA, Haplin D, Charlton H, Gordon S: The mononuclear phagocyte system of the mouse defined by immunohistochemical localization of antigen F4/80: Macrophages of endocrine organs. Proc Natl Acad Sci USA 1984;81:4171–4177.
5 Dinarello CA: Biology of interleukin-1. FASEB J 1988;2:108–115.
6 Dunn AJ: Interleukin-1 as a stimulator of hormone secretion. Prog Neuro Endocrin Immunol 1990;3:26–34.
7 Farrar WL, Kilian PL, Ruff MR, Hill JM, Pert CB: Visualization and characterization of interleukin-1 receptors in brain. J Immunol 1987;139:459–463.
8 Marguette C, Ban E, Fillon G, Haour F: Receptors for interleukin 1, 2 and 6 (IL-1α and β, IL-2, IL-6) in mouse, rat and human pituitary. Neuroendocrinology 1990;52: 48.

9 Farrar W, Hill J, Hael-Bellan A, Vinocour M: The immune logical brain. Immunol Rev 1987;100:361–378.
10 Lechan RM, Toni R, Clark BD, Cannon JG, Shaw AR, Dinarello CA, Reichlin S: Immunoreactive interleukin-1β localization in the rat forebrain. Brain Res 1990; 514:135–140.
11 Woloski BMRNJ, Smith EM, Meyer WJ, Fuller GM, Blalock JE: Corticotropin-releasing activity of monokines. Science 1985;230:1035–1037.
12 Sapolsky R, Rivier C, Yamamoto G, Plotsky P, Vale W: Interleukin-1 stimulates the secretion of hypothalamic corticotropin-releasing factor. Science 1987;238:522–524.
13 Berkenbosch F, Oers J van, Rey A del, Tilders F, Besedowsky H: Corticotropin-releasing factor-producing neurons in the rat activated by interleukin-1. Science 1987;238:524–526.
14 Bernton EW, Beach JE, Holaday JW, Smallridge RC, Fein HC: Release of multiple hormones by a direct action of interleukin-1 on pituitary cells. Science 1987;238: 519–521.
15 Uehara A, Gottschall PE, Dahl RR, Arimura A: Stimulation of ACTH release by human interleukin-1β, but not by interleukin-1α, in conscious, freely-moving rats. Biochem Biophys Res Commun 1987;146:1286–1290.
16 Brown SL, Smith LR, Blalock JE: Interleukin-1 and interleukion-2 enhance proopiomelanocortin gene expression in pituitary cells. J Immunol 1987;139:3181–3183.
17 Fukata J, Isui T, Naitoh Y, Nakai Y, Imura H: Effects of recombinant human interleukin-1α, -β, 2 and 6 on ACTH synthesis and release in the mouse pituitary tumor cell line AtT-20. J Endocrinol 1989;122:33–39.
18 Uehara A, Gillis S, Arimura A: Effects of interleukin-1 on hormone release from normal rat pituitary cells in primary culture. Neuroendocrinology 1987;45:343–347.
19 Malarkey WB, Zvara BJ: Interleukin-1β and other cytokines stimulate adrenocorticotropin release from cultured pituitary cells by patients with Cushing's disease. J Clin Endocrinol Metab 1989;69:196–199.
20 Kehrer P, Turnill D, Dayer JM, Muller AF, Gaillard RC: Human recombinant interleukin-1β and -α, but not recombinant tumor necrosis factor α stimulate ACTH release from rat anterior pituitary cells in vitro in a prostaglandin E_2 and cAMP independent manner. Neuroendocrinology 1988;48:160–166.
21 Fagarasan MO, Eskay R, Axelrod J: Interleukin-1 potentiates the secretion of β-endorphin induced by secretagogues in a mouse pituitary cell line (AtT-20). Proc Natl Acad Sci USA 1989;86:2070–2073.
22 Fagarasan MO, Aiello F, Muegge K, Durum S, Axelrod J: Interleukin-1 induces β-endorphin secretion via FOS and Jun in AtT-20 pituitary cells. Proc Natl Acad Sci USA 1990;87:7871–7874.
23 Boyle M, Yamamoto G, Chen M, Rivier J, Vale W: Interleukin-1 prevents loss of corticotropic responsiveness to β-adrenergic stimulation in vitro. Proc Natl Acad Sci USA 1988;85:5556–5560.
24 Schettini G: Interleukin-1 in the neuroendocrine system: From gene to function. Prog Neuro Endocrin Immunol 1990;3:157–166.
25 Besedovsky H, Rey A del, Sorkin E, Dinarello CA: Immunoregulatory feedback between interleukin-1 and glucocorticoid hormones. Science 1986;233:652–654.
26 Ovadia H, Abramsky O, Barak V, Conforti N, Saphier D, Weidenfeld J: Effect of interleukin-1 on adrenocortical activity in intact and hypothalamic deafferentated male rats. Exp Brain Res 1989;76:246–249.

27 Navarra P, Tsagarakis S, Faria MS, Rees LH, Besser GM, Grossman AB: Interleukins-1 and -6 stimulate the release of corticotropin-releasing hormone-41 from rat hypothalamus in vitro via the eicosanoid cyclooxygenase pathway. Endocrinology 1990;128:37–44.
28 Bartfai T, Andersson C, Bristulf J, Schultzberg M, Svenson S: Interleukin-1 in the noradrenergic chromaffin cells in the rat adrenal medulla. Ann NY Acad Sci 1990; 594:207–213.
29 Winter JSD, Gow KW, Perry YS, Greenberg AH: A stimulatory effect of interleukin-1 on adrenocortical cortisol secretion mediated by prostaglandins. Endocrinology 1990;127:1904–1909.
30 Dyck DG, Janz L, Osachuk TAG, Falk J, Labinsky J, Greenberg AH: The Pavlovian conditioning of IL-1-induced glucocorticoid secretion. Brain Behav Immun 1990;4: 93–104.
31 Fukuoka M, Taii S, Yasuda K, Takakura K, Mori T: Inhibitory effects of interleukin-1 on luteinizing hormone-stimulated adenosine 3′, 5′-monophosphate accumulation by cultured porcine granulosa cells. Endocrinology 1989;125:136–143.
32 Fukuoka M, Yasuda K, Taii S, Takakura K: Interleukin-1 stimulates growth and inhibits progesterone secretion in cultures of porcine granulosa cells. Endocrinology 1989;124:884–890.
33 Yagel S, Lala PK, Powell WA, Casper RF: Interleukin-1 stimulates human chorionic gonadotropin secretion by first trimester human trophoblast. J Clin Endocrinol Metab 1989;68:992–995.
34 Croy BA, Malashenko B, Poterski R, Yamashiro S, Summerlee AJS: Intracerebroventricular administration of interleukin-1β induces postimplantation pregnancy failure in mice. Prog Neuro Endocrin Immunol 1990;3:242–250.
35 Sandler S, Andersson A, Hellerstrom C: Inhibitory effects of interleukin-1 on insulin secretion, insulin biosynthesis, and oxidative metabolism of isolated rat pancreatic islets, Endocrinology 1987;121:1424–1431.
36 Breder CD, Dinarello CA, Saper CB: Interleukin-1 immunoreactive innervation of the human hypothalamus. Science 1988;240:321–324.
37 Stefano GB, Smith EM, Hughes TK: Opioid induction of immunoreactive interleukin-1 in *Mytilus edulis* and human immunocytes: an interleukin-1 substance in invertebrate neural tissue. J Neuroimmunol 1991;32:29–34.
38 Haour FG, Ban EM, Milon GM, Baran D, Fillion GM: Brain interleukin-1 receptors: Characterization and modulation after lipopolysaccharide injection. Prog Neuro Endocrin Immunol 1990;3:196–204.
39 Takao T, Tracey DE, Mitchell WM, DeSouza EB: Interleukin-1 receptors in mouse brain: Characterization and neuronal localization. Endocrinology 1990;127:3070–3078.
40 Blatteis CM: Functional similarities between somnogens and pyrogens: Fortuity or grand design? In Inoue S, Krueger JM (eds): Endogenous Sleep Factors. The Hague, SPB Academic Publishing, 1990, pp 249–256.
41 Krueger JM, Toth LA, Johannsen L, Obal F, Opp M, Cady AB: Enhancement of mammalian sleep by immune modulators; in Hadden J, Spreafico F, Yamamura Y, Austen KF, Dukor P, Masek K (eds): Advances in Immunopharmacology. Oxford, Pergamon, 1989, vol 4, pp 55–63.
42 Besedovsky HO, Sorkin E, Felix D, Haas H: Hypothalamic changes during the immune response. Eur J Immunol 1977;7:323–325.

43 Besedovsky HO, del Rey A, Sorkin E, Da Prada M, Burri R, Honegger C: The immune response evokes changes in brain noradrenergic neurons. Science 1983;221: 564–566.
44 Akahoshi T, Oppenheim JJ, Matsushima K: Interleukin-1 stimulates its own receptor expression on human fibroblasts through the endogenous production of prostaglandin(s). J Clin Invest 1988;82:1219–1224.
45 Burch RM, White MF, Connor JR: Interleukin-1 stimulates prostaglandin synthesis and cyclic AMP accumulation in Swiss 3T3 fibroblasts: interactions between two second messenger systems. J Cell Physiol 1989;139:29–33.
46 Murakami N, Watanabe T: Activation of ACTH release is mediated by the same molecule as the final mediator, PGE_2, of febrile response in rats. Brain Res 1989;478: 171–174.
47 Hughes TK, Chin R, Smith EM, Leung MK, Stefano GB: Similarities of signal systems in vertebrates and invertebrates: detection, action and interactions of immunoreactive monokines in the mussel, *Mytilus edulis*. Adv Neuroimmunol 1991;1: 59–69.
48 Mosmann TR, Coffman RL: TH_1 and TH_2: different patterns of lymphokine secretion lead to different functional properties. Annu Rev Immunol 1989;7:145–173.
49 Besedovsky HO, del Rey A, Sorkin E: Lymphokine-containing supernatants from Con A-stimulated cells increase corticosterone blood levels. J Immunol 1981;126: 385–387.
50 Lotze MT, Frana LW, Sharrow SO, Robb RJ, Rosenberg SA: In vivo administration of purified human interleukin-2. J Immunol 1985;134:157–166.
51 Rosenberg SA, Lotze MT, Muul LM, Chang AE, Avis FP, Leitman S, Linehan WM, Robertson CN, Lee RE. Rubin JT, Seipp CA, Simpson CG, White DE: A progress report on the treatment of 157 patients with advanced cancer using lymphokine-activated killer cells and interleukin-2 or high dose interleukin-2 alone. N Engl J Med 1987;316:889–897.
52 West WH, Tauer KW, Yannelli JR, Marshall GD, Orr DW, Thurman GB, Oldham RK: Constant-infusion recombinant interleukin-2 in adoptive immunotherapy of advanced cancer. N Engl J Med 1987;316:898–905.
53 Roosth J, Pollard RB, Brown SL, Meyer WJ: Cortisol stimulation by recombinant interferon-α_2. J Neuroimmunol 1986;12:311–316.
54 Rinehart JJ, Young D, Laforge J, Colburn D, Neidhart J: Phase I/II trial of recombinant γ-interferon in patients with renal cell carcinoma. Immunologic and biologic effects. J Biol Response Mod 1987;6:302–312.
55 Gutterman JU, Fine S, Quesada J, Horning SJ, Levine JF, Alexanian R, Bernhardt L, Kramer M, Spiegel H, Colburn W, Trown P, Merigan T, Dziewanowski Z: Recombinant leukocyte α-interferon: pharmacokinetics, single-dose tolerance, and biologic effects in cancer patients. Ann Intern Med 1982;96:549–556.
56 Karanth S, McCann SM: Anterior pituitary hormone control by interleukin-2. Proc Natl Acad Sci USA 1991;88:2961–2965.
57 Wolvekamp MCJ, Marquet RL: Interleukin-6: historical background, genetics and biological significance. Immunol Lett 1990;24:1–9.
58 Spangelo BL, MacLeod RM: Regulation of the acute phase response and neuroendocrine function by interleukin-6. Prog Neuro Endocrin Immunol 1990;3:167–175.
59 Synder DS, Unanue ER: Corticosteroids inhibit murine macrophage Ia expression and interleukin-1 production. J Immunol 1982;129:1803–1805.

60 Naitoh Y, Fukata J, Tominaga T, Nakaai Y, Tamai S, Mori K, Imura H: Interleukin-6 stimulates the secretion of adrenocorticotropic hormone in conscious, freely moving rats. Biochem Biophys Res Commun 1988;155:1459–1463.
61 Frei K, Malipiero UV, Leist TP, Zinkernagel RM, Schwab ME, Fontana A: On the cellular source and function of interleukin-6 produced in the central nervous system in viral diseases. Eur J Immunol 1989;19:689–694.
62 Spangelo BL, Judd AM, MacLeod RM, Goodman DW, Isakson PC: Endotoxin-induced release of interleukin-6 from rat medial basal hypothalami. Endocrinology 1990;127:1779–1785.
63 Spangelo BL, MacLeod RM, Isakson PC: Production of interleukin-6 by anterior pituitary cells in vitro. Endocrinology 1990;126:582–586.
64 De Simoni MG, Sironi M, De Luigi A, Manfridi A, Mantovani A, Ghezzi P: Intracerebroventricular injection of interleukin-1 induces high circulating levels of interleukin-6. J Exp Med 1990;171:1773–1778.
65 Yamaguchi M, Yoshimoto Y, Koike K, Matsuzaki N, Miyake A, Tanizawa O: Possible involvement of interleukin-6 in the mechanism of hypothalamic GnRH secretion induced by estradiol. Neuroendocrinology, in press.
66 Beutler B, Cerami A: The biology of cachectin/TNF-α primary mediator of the host response. Annu Rev Immunol 1989;7:625–655.
67 Cambell IL, Iscaro A, Harrison LC: INF-γ and tumor necrosis factor-α cytotoxicity to murine islets of Langerhans. J Immunol 1988;141:2325–2329.
68 Satoh J, Seino H, Shintani S, Tanaka S-I, Ohteki T, Masuda T, Nobunaga T, Toyota T: Inhibition of type 1 diabetes in BB rats with recombinant human tumor necrosis factor-α. J Immunol 1990;145:1395–1399.
69 Jiang Z, Woda BA: Cytokine gene expression in the islets of the diabetic biobreeding/Worcester rat. J Immunol 1991;146:2990–2994.
70 Walton PE, Cronin MJ: Tumor necrosis factor-α inhibits growth hormone secretion from cultured anterior pituitary cells. Endocrinology 1989;125:925–929.
71 Chung IY, Benveniste EN: Tumor necrosis factor-α production by astrocytes, induction by lipopolysaccharide, IFN-γ, and IL-1β. J Immunol 1990;144:2999–3007.
72 Pang X-P, Hershman JM, Mirell CJ, Pekary AE: Impairment of hypothalamic-pituitary-thyroid function in rats treated with human recombinant tumor necrosis factor-α (cachectin). Endocrinology 1989;125:76–84.
73 Spengler RN, Allen RM, Remick DG, Strieter RM, Kunkel SL: Stimulation of α-adrenergic receptor augments the production of macrophage-derived tumor necrosis factor. J Immunol 1990;145:1430–1434.
74 Yelavarthi KK, Chen H, Yang Y, Cowley BD, Fishback JL, Hunt JS: Tumor necrosis factor-α mRNA and protein in rat uterine and placental cells. J Immunol 1991;146:3840–3848.
75 Blalock JE, Stanton JD: Common pathways of interferon and hormonal action. Nature (Lond) 1980;283:406–408.
76 Blalock JE, Harp C: Interferon and adrenocorticotropic hormone induction of steroidogenesis, melanogenesis and antiviral activity. Arch Virol 1981;67:45–49.
77 Chany C, Mathieu D, Gregoire A: Induction of γ-ketosteroid synthesis by interferon in mouse adrenal tumor cell cultures. J Gen Virol 1980;50:447–450.
78 Lawrence TS, Beers WH, Gilula NB: Transmission of hormonal stimulation by cell-to-cell communication. Nature (Lond) 1978;272:501–506.

79 Cavet MC, Gresser I: Interferon enhances the excitability of cultured neurons. Nature (Lond) 1979;278:558–560.
80 Blalock JE, Smith EM: Human leukocyte interferon (HuIFN-α). Potent endorphin-like opioid activity. Biochem Biophys Res Commun 1981;101:472–478.
81 Dafny N: Interferon: a candidate as the endogenous substance preventing tolerance and dependence to brain opioids. Prog Neuropsychopharmacol Biol Psychiatry 1984;8:351–357.
82 Reyes-Vasquez C, Prieto-Gomez B, Georgiades JA, Dafny N: Alpha- and gamma-interferons' effects on cortical and hippocampal neurons. Microiontophoretic application and single cell recording. Int J Neurosci 1984;25:113–121.
83 Dafny N, Prieto-Gomez B, Reyes-Vasquez C: Does the immune system communicate with the central nervous system? J Neuroimmunol 1985;9:1–12.
84 Dafny N, Reyes-Vasquez C: Three different types of α-interferons alter naloxone-induced abstinence in morphine-addicted rats. Immunol Pharmacol 1985;9:13–17.
85 D'Urso R, Falaschi P, Canfalone G, Carusi E, Proietti A, Barnaba V, Balsano F: Neuroendocrine effects of recombinant α-interferon administration in humans. Prog Neuro Endocrin Immunol 1991;4:20–25.
86 Billiau A, Denef C: IL-1 interferon-γ inhibits stimulated adrenocorticotropin, prolactin, and growth hormone secretion in normal rat anterior pituitary cell cultures. Endocrinology 1990;126:2919–2926.
87 Munck A, Guyre PM, Holbrook NJ: Physiological functions of glucocorticoids in stress and their relation to pharmacological actions. Endocrinol Rev 1984;5:24–44.
88 Roth JA, Frank DE: Recombinant bovine interferon-γ as an immunomodulator in dexamethasone-treated and nontreated cattle. J Interferon Res 1989;9:143–151.
89 Gonzalez MC, Riedel M, Rettori V, Yu WH, McCann SM: Effect of recombinant human γ-interferon on the release of anterior pituitary hormones. Prog Neuro Endocrin Immunol 1990;3:49–54.
90 Kuriyama K, Hori T, Mori T, Nakashima T: Actions of interferon-α and interleukin-1β on the glucose-responsive neurons in the ventromedial hypothalamus. Brain Res Bull 1990;24:803–810.
91 Goldstein D, Bushmeyer SM, Witt PL, Jordan VC, Borden EC: Effects of type I and II interferons on cultured human breast cells: interaction with estrogen receptors and with tamoxifen. Cancer Res 189;49:2698–2702.
92 Triozzi PL, Avery KB, Abou-Issa HM, Chou TC: Combined effects of interferon and steroid hormones on 2′,5′-oligoadenylate synthetase activity in chronic lymphocytic leukemia cells. Leuk Res 1989;13:437–443.
93 Almawi WY, Lipman ML, Stevens AC, Zanker B, Hadro ET, Strom TB: Abrogation of glucocorticosteroid-mediated inhibition of T cell proliferation by the synergistic action of IL-1, IL-6, and IFN-γ. J Immunol 1991;146:3523–3527.
94 Kavelaars A, Ballieux RE, Heijnen CJ: The role of interleukin-1 in the CRF- and AVP-induced secretion of ir-β-endorphin by human peripheral blood mononuclear cells. J Immunol 1989;142:2338–2342.
95 Judd AM, Spangelo BL, MacLeod RM: Rat adrenal zona glomerulosa cells produce interleukin-6. Prog Neuro Endocrin Immunol 1990;3:282–292.

Eric M. Smith, PhD, Department of Psychiatry and Behavioral Sciences,
University of Texas Medical Branch, Galveston, TX 77550 (USA)

Blalock JE (ed): Neuroimmunoendocrinology, 2nd rev ed.
Chem Immunol. Basel, Karger, 1992, vol 52, pp 170–190

Signaling Pathways of the Neuroendocrine-Immune Network[1]

Thomas L. Roszman, William H. Brooks

Department of Microbiology and Immunology,
University of Kentucky Medical Center, Lexington, Ky., USA

Introduction

Evidence continues to accumulate indicating that interaction between the neuroendocrine and immune systems has potential biological significance. Similar to the well-described hormonal loops within the endocrine system, neuroendocrine-immune interactions would be anticipated to involve feedback and feedforward loops thus maintaining homeostasis. These bidirectional loops occur as a consequence of neurohormones acting on cells of the immune system while cytokines or antibodies secreted by lymphocytes and macrophages in turn influence neuroendocrine function. One example of this relationship exists between adrenal corticotropin hormone (ACTH) and interleukin-1 (IL-1). Thus, IL-1 produced by macrophages can influence the production of ACTH by direct modulation of ACTH-producing cells in the pituitary gland [12, 23] or indirectly through the release of corticotropin-releasing hormone from the hypothalamus [10, 82]. The increase in ACTH secretion results in enhanced synthesis of corticosteroids which can in turn decrease IL-1 production by macrophages [15]. It is important to note that such interactions are not only inhibitory but facilitatory as well, fulfilling important teleologic considerations. Furthermore, examination of data related to neuroendocrine-immune interactions reveals that under physiologic conditions alterations occurring in either system generally are small both in magnitude and duration, thereby suggesting the existence of a homeostatic relationship between these two systems. The ability of one system to dominate the other would lead to chaos. Similar to all biological regulatory

[1] Supported in part by US Public Health Grant NS17423.

systems this interaction should exhibit the ability to generate an amplified response to low levels of stimuli as well as manifest the ability to adapt to prolonged high levels of stimuli further underscoring the function of these systems as maintaining homeostasis within the host.

The focus of this review is to amplify these concepts and pose questions concerning the relationship between the neuroendocrine and immune systems with emphasis on possible mechanisms of how neurohormones modulate lymphocyte function.

Can the Pituitary Gland Modulate the Immune System?

It is reasonable to hypothesize that if the pituitary gland can influence the immune system, then its absence or removal should result in detectable alterations in immunity. Indeed, there are a number of reports, using different approaches, which support this contention. Chief among these are those that have focused on the immune potential of hypopituitary Snell-Bagg mice. These mice generally exhibit a decrease in the number of cells in the thymus, bone marrow and peripheral lymphoid tissue [3, 4, 41–44, 72] with accompanying impairment in both humoral and cell-mediated immune responses [3, 4, 41–44, 72], although there are reports indicating that the immune system of these animals is normal [40, 84]. Overall these data strongly suggest that pituitary-derived hormones play an important role in the development and maturation of the immune system and/or the antigen-induced events responsible for initiation of an immune response. Immunologic studies of hypophysectomized mice and rats have yielded results similar to those obtained with hypopituitary dwarf mice. The majority of these studies, but not all [39, 56, 89, 90], demonstrate that hypophysectomized animals exhibit impaired humoral [29, 46, 67, 71] and cell-mediated immunity [29, 37, 71, 75, 83].

In general there have been few, if any, attempts to examine at the mechanistic level the effects of hypophysectomy on immune function. However, it is clear that a hypophysectomized animal can be immunologically reconstituted by a number of relatively straightforward maneuvers including the administration of either growth hormone or prolactin, as well as ectopic pituitary grafts [7, 9]. The immunomodulatory role of these hormones is further substantiated by the observation that implantation of GH_3 pituitary tumor cells into aged syngeneic rats results in cellular and functional restoration of the thymus to levels comparable to that of 3-month-old animals [59]. Although direct proof is lacking, the presumption is that this restoration results from the production of prolactin and growth hormone by the GH_3 tumor cells. We have demonstrated that normal C57BL/6 mice given pituitary grafts under the kidney capsule, which have a 2- to 3-fold sustained increase in prolactin, have markedly greater primary antibody responses to sheep red blood cells (SRBC) as compared to normal values [34]. Interestingly, placement of pituitary grafts under the kidney capsule of 24-month-old mice does not result in cellular or functional restoration of the thymus to comparable levels observed in young animals, suggesting that prolactin may not have a role in this process [33].

There is other evidence suggesting that the pituitary gland has a key role in modulating immune function. We initiated a series of experiments directed at confirming earlier reports that anterior hypothalamic ablation is associated with diminished immune reactivity [87]. In these studies we demonstrated that electrolytic lesioning of the anterior hypothalamic areas of rats induces a decrease in the numbers of nucleated spleen cells and thymocytes as compared to control lesioned and normal animals [36]. This modulation extends also to mitogen [36] and antigen-driven lymphocyte proliferation [78] as well as natural killer cell function [37]. The observed diminished immune reactivity was short-lived; returning to normal within 14 days. Exploration of other areas of the brain indicated that in addition to areas that are associated with down-regulation of immunity, there are specific areas (hippocampus and amygdaloid complex) that facilitate lymphocyte reactivity with a markedly enhanced proliferative response [22]. These effects also disappeared by 14 days after lesioning. The acuteness of the effects is attributed to the remarkable plasticity of the CNS and the subsequent re-establishment of dendritic-axonal networks [47].

From these experiments demonstrating that changes in the CNS function can be correlated with alterations in immune reactivity, we began the search for the functional link between the CNS and mediators of immunity. We have shown that modulation of immunity by neural lesions is not related to changes in corticosterone [36]; yet certain pituitary-derived hormones are influenced by anterior hypothalamic lesions and these are capable of impairing cellular activation. To test this hypothesis, neurally induced changes in lymphocyte responsiveness were assessed before and after hypophysectomy [32]. The results of these investigations suggest that the CNS is capable of modulating splenocyte blastogenesis and number via the endocrine system because pituitary removal abrogates both the inhibitory as well as the facilitatory immunologic effects induced by neural lesions. In a similar system [35] we have also examined the effects of 'chemical' neural lesions produced by the cisterna magna injection of 6-hydroxydopamine (6-OHDA). Treatment of animals with 6-OHDA 48 h prior to immunization results in a profound suppression of the primary IgM and IgG antibody response which is not the result of increased levels of corticosterone. However, if 6-OHDA is injected into hypophysectomized mice 48 h prior to immunization, the primary antibody response is not impaired compared to the response of noncatecholamine-depleted hypophysectomized mice [31]. These results strongly suggest that modulation of the immune response as a consequence of central catecholamine depletion is emanating from the pituitary gland.

Collectively, these data indicate that the pituitary gland via the hormones it elaborates has a decided modulatory effect on immune function. Precisely how these hormones can modulate immunity remains to be determined. In fact there are few, if any, experiments which probe the mechanism of action of hypophysectomy on immune function. For example, does hypophysectomy influence immune function by inducing alterations in the balance between helper-suppressor circuits, lymphocyte maturation, antigen processing and presentation or cell trafficking? It is clear that the immunosuppressive effects of hypophysectomy can be overcome by either growth hormone or prolactin [7, 9] thus possibly providing an important clue as to the mechanisms responsible for this effect. While hypophysectomy does establish the importance of neurohormones in modulating immune function, it may be difficult to use this approach to decipher the precise mechanism. Removal of the pituitary gland causes a compounding number of changes in the normal physiology of the animal thereby making it difficult to determine precisely what factor or combination of factors are responsible for the observed alterations in immune function.

Are There Receptors for Neurohormones Present on Cells of the Immune System?

Data implicating the pituitary gland in modulating antigen independent and dependent maturation of the immune system suggest that neurohormones have a vital role in this process. A pivotal concept in neuroendocrine-lymphocyte interactions is the presence of receptors in or on cells of the immune system. Therefore, if neurohormones are to modulate immune function there must be external and/or internal receptors for these peptides in order for the cells of the immune system to receive and subsequently

Table 1. Receptors for pituitary hormones on cells of the immune system

Hormone	Source of cells	Reference
Growth hormone	human lymphocyte cell line IM-9	64, 65
	calf and mouse thymocytes	2
Prolactin	human T cells, B cells and monocytes	80
	human lymphocytes	6, 81
ACTH	mouse splenic mononuclear cells	54
	human mononuclear lymphocytes	86
	rat T and B cells	27
β-Endorphin	cultured human lymphocytes	48
	human peripheral blood lymphocytes	69
Thyrotropin	human monocytes and polymorphonuclear leukocytes	26
	human monocytes, natural killer cells and activated B cells	30

interpret the putative signal and, there is evidence for such receptors on these cells. Because the results of these studies have been reviewed often and recently [8, 14, 16, 17, 19–21, 25, 58, 68, 73, 74, 85] only salient features required to illustrate certain points will be highlighted. Some of the better characterized receptors for pituitary-derived hormones reported to be detectable on lymphoid and myeloid cells are listed in table 1. Although these data provide presumptive evidence for the presence of neurohormone receptors, there is clearly a need for confirmatory studies using antibodies and molecular probes. Accordingly, initial studies with prolactin have demonstrated the presence of the prolactin receptor in a murine T-helper cloned cell line employing both antibodies and a cDNA-specific probe for the receptor [28]. Interestingly, these studies also demonstrate that the level of the prolactin receptor increases following stimulation of the cells with either interleukin-2 (IL-2) or concanavalin A. These studies illustrate a number of important corollaries concerning the presence of neurohormone receptors on lymphocytes. For example, these receptors may be differentially expressed on certain subsets of T cells; not all lymphocytes may possess neurohormone receptors and hence would be refractory to hormonal modulation. Moreover, expression of neurohormone receptors may be autoregulated during activation of the lymphocyte; an increase in receptor expression might amplify the effects of neurohormones on cellular function whereas a diminution or loss of receptors would be expected to have an opposing effect. Finally, although the utility of employing cloned T-cell lines or lymphoblastoid cells has obvious advantages, it is important to realize that the presence of neurohormone receptors on these 'transformed' cells does not guarantee their presence on normal lymphocytes.

Can Neurohormones Influence the Immune System?

The biological significance of the presence of neurohormone receptors on the cells of the immune system is supported by the in vivo and in vitro evidence attesting to the ability of neurohormones to modify cellular behavior. Because there are a number of comprehensive recent reviews regarding the effects of neurohormones on immune function [7–9, 11, 14, 17, 19–21, 25, 38, 55, 58, 85], we will restrict our comments to a general overview of the phenomenology revealed by these studies. One readily apparent observation drawn from these studies is that neurohormones have broad immunomodulatory effects on various types of cells involved in the immune response. For example, β-endorphin alters the function of lymphocytes, natural killer cells, macrophages, polymorphonuclear cells and mast cells [85]. Moreover the effects of a specific neurohormone on cell function can either be facilitatory or inhibitory [11, 38, 58, 85]. This appears to be a general pattern for most neurohormones. Overall, these data imply that the various and diverse cells encompassing the immune system have receptors for neurohormones and that the biochemical and molecular signals elicited in these cells as a consequence of ligand-receptor interactions are differentially encoded and interpreted. Unfortunately, there is currently a paucity of data to substantiate this hypothesis; more research is required to clarify these interesting but confounding observations. This will not be a trivial undertaking considering the complexity of the immune system and its cells, compounded by the fact that many neurohormones exert only modest alterations in immune function.

How Do Neurohormones Modulate Lymphocyte Function?

Given the evidence that neurohormones modulate immunity, the challenge now is to understand the mechanism(s) of action. In discussing potential mechanisms responsible for this form of neuroimmunomodulation, we will restrict our comments to acquired immunity involving predominantly the T cell. It should be emphasized that this does not mean that innate immunity is not modulated by neurohormones; for example, growth hormone has a profound facilitatory effect on phorbol ester-induced superoxide production by polymorphonuclear cells [45]. Although a teleologic argument can be presented for neurohormone modulation of natural immunity, much more evidence is required for support.

In discussing the modulation of T-cell function by pituitary-derived hormones, it is useful to recall briefly the heterogeneity of these cells. There are two major types of T cells: immature and mature; the latter subset being further categorized in a general way as naive, memory or effector cells. Moreover, it is important to appreciate that these lymphocytes can be either resting (G_0 of the cell cycle) or activated (other than G_0). The ability of activated T cells to up-regulate hormone receptor expression suggests that the modulatory effects of neurohormones may be more readily apparent in activated T cells. Thus, it can be hypothesized that the susceptibility of T cells to neurohormonal regulation is determined by its state of activation and/or differentiation. Because T cells are involved in a variety of activities, e.g. thymic education, trafficking and effector functions, etc., the spectrum of phenomena potentially modulatable by neurohormones is immense. To date, these questions have not been examined in any detail.

Perhaps the best studied example of neurohormone modulation of immunity is T-cell activation. Generally two questions are posed: how is T-cell function modulated, and what biochemical and molecular mechanisms are involved? The former receives far more attention than the latter (see above). In designing experiments to analyze the second question, those key elements involved in T-cell activation, e.g. antigen, cell contact and lymphokines, must be considered. The initial events in this process involve the binding of antigen to the TCR on T cells presented in the context of either class I or II major histocompatibility antigens on accessory cells. The biochemical signals elicited by this interaction are subsequently combined with those resulting from the binding of adhesion molecules on T cells with their appropriate ligands as well as those initiated by such lymphokines as IL-1 and IL-2 binding to their receptors. This results in a cascade of integrated biochemical and molecular events in the resting T cell leading ultimately to proliferation [1]. Although the details of T-cell activation are much more complex than portrayed in this simplistic scenario, this model of T-cell activation and proliferation serves to focus on the possible sequence of events that are amenable to neurohormonal modulation.

Presumably neurohormones modulate the function of T cells by binding to their specific receptor and eliciting induction of second messengers. The evidence identifying which second messengers are elicited by various neurohormones is sparse, but there are clues as to those most probably involved in neuroendocrine-immune modulation. For example, stimulation of human peripheral blood T cells with β-endorphin modulates the phorbol ester-induced phosphorylation of the γ chain of the CD3 complex associated with the TCR [57]. Although the mechanism responsible for this remains to be identified, the data suggest that activation of protein kinase C (PKC) is responsible. Preincubation of the Jurket T-cell line with β-endorphin induces the release of intracellular calcium (Ca^{2+}) and an increase in c-myc expression via a potassium channel [53]. Others have demonstrated that β-endorphin can increase $^{45}Ca^{2+}$ uptake into concanavalin A-stimulated rat thymocytes [49]. This same group has also demonstrated that β-endorphin can partially overcome the prostaglandin E_2 (PGE_2) suppression of the phytohemagglutinin responsiveness of rat lymph node cells [50]. A direct effect of β-endorphin on adenylyl cyclase is unlikely as this peptide does not inhibit either the forskolin or cholera toxin-induced suppression of the PHA response of the lymph node cells. These results do not eliminate the possible involvement of the modulation of cAMP and protein kinase A (PKA) by β-endorphin. Thus, the possibility remains that β-endorphin modulates the PGE_2 receptor, cAMP phosphodiesterase activity, a G-inhibitory protein, and/or PKA. In part, this question can be answered by determining the level of PGE_2-induced cAMP in these β-endorphin-treated lymphocytes in the presence or absence of a cAMP phosphodiesterase inhibitor. There is also evidence to suggest that prolactin activates nuclear PKC in lymphocytes [79] and that ACTH induces cAMP synthesis in these cells [27]. Review of existing data provides evidence for the presence of functional receptors for neurohormones on lymphocytes capable of modulating their function by eliciting a variety of second messengers. The mechanisms by which hormonally generated signaling modify T-cell second messenger systems and how the modulations interfere with lymphocyte activation and/or proliferation remain elusive.

Neurohormone-Lymphocyte Interactions: Signaling Models

In considering possible approaches to unraveling the mechanisms of neuroendocrine-immune modulation we have found it useful to develop

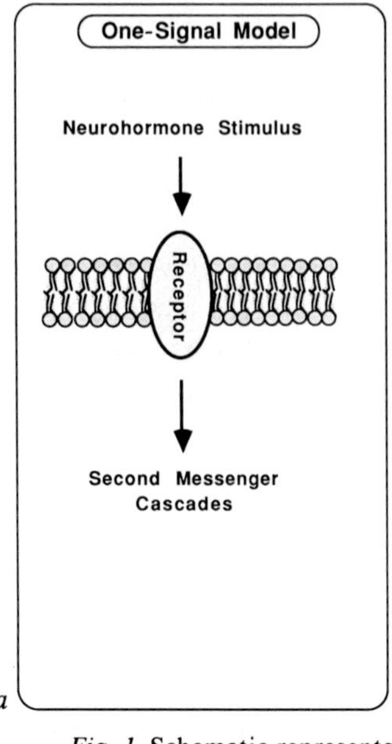

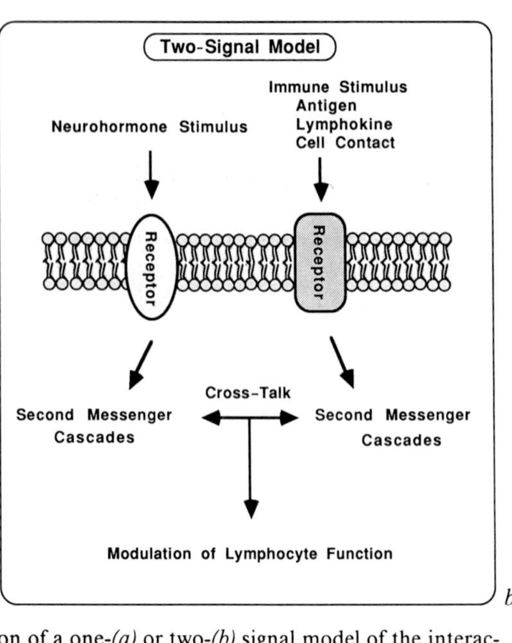

Fig. 1. Schematic representation of a one-*(a)* or two-*(b)* signal model of the interaction of neurohormones and immune stimuli with their respective receptors on lymphocytes. These models assume that the lymphocytes are in G_0 of the cell cycle (resting). This representation shows only the early events that occur as a result of appropriate ligand-receptor interaction. See text for a detailed discussion of these models.

models which may serve as a template for subsequent inquiry. These models represent an extension of the tenets of neurohormone-lymphocyte interaction as discussed previously and are based on the integrated biological systems of signal amplification and the cascade of molecular and biochemical events, which result in modulation of function and adaptability. To achieve this end we propose two minimal models, both amenable to experimentation, that may prove useful in understanding how neurohormones physiologically alter T cells and their activation by immune stimuli. The first is a one-signal model which considers the consequences of a neurohormone interacting with its receptor on a resting T cell (fig. 1a). The other is a two-signal model involving the binding of a neurohormone to its receptor on T cells concurrent with the T cell receiving another signal from antigen interacting with the TCR (fig. 1b).

As previously discussed, the binding of a neurohormone to its receptor on T cells results in the generation of a biochemical wave of second messengers through the cell. In a one-signal model, the level of second messengers generated and contingent biochemical and molecular events within the T cell is dependent on the concentration of neurohormone. Thus, a rise in the concentration of hormone above its normal steady-state level will induce magnitude amplification [60] dependent upon the number of molecules of the neurohormone bound to its receptor on T cells. For example, if 500 molecules of neurohormone bind to their receptors on T cells, 10^4 molecules of a particular second messenger may be generated. Well-known biologic events that increase neurohormone levels include stress, estrus or pregnancy as well as the production of neurohormones by lymphoid cells [18]. While interaction of the neurohormone with T cells results in a diverse number of biochemical and molecular alterations in the cell, this interaction does not induce competence and progression. In this one-signal model the T cell will undergo adaptation employing a number of compensatory mechanisms. To illustrate this, ACTH can serve as a useful example. As a consequence of stress the level of circulating ACTH is increased and will bind to its receptor on lymphocytes [27, 54, 86] eliciting the second messenger cAMP [27]. A resting T cell receiving no other inputs represents an example of the one-signal model. The cell initially changes biochemically and molecularly as a result of interacting with ACTH but does not undergo adaptation. This adaptation occurs through a number of well-known mechanisms including receptor down-regulation, uncoupling of adenylyl cyclase and increasing cAMP phosphodiesterase. With time, and as the ACTH returns to baseline values, the cell completes its adaptation, returns to its resting state and again becomes receptive to this peptide.

A two-signal model raises a number of complex but fundamentally important questions relating to the modulation of lymphocyte function. As the name indicates, the lymphocyte receives one signal from neurohormone binding to its receptor and the other from an immune stimulus, e.g. antigen, cell contact or lymphokine. Assuming both signals arrive at or about the same time, the lymphocyte generates two similar or dissimilar sets of second messengers that must be decoded and integrated into either previous or subsequent biochemical and molecular events. Ultimately, this is reflected in modulation of lymphocyte function. A specific example of this is the inhibitory effect of ACTH on T-cell function. As previously discussed, binding of ACTH to its receptor on lymphocytes results in increased synthesis of cAMP. Assuming that ACTH binds to a T cell undergoing stimulation with an antigen, two distinctly different sets of second messengers will be induced. Antigen binding to the TCR complex activates phospholipase C (PLC) resulting in the initiation of the phosphatidylinositol cycle yielding two important second messengers, calcium (Ca^{2+}) and PKC [1]. Thus, phosphatidylinositol bisphosphate is hydrolyzed by PLC to yield inositol trisphosphate (IP_3) and diacylglycerol (DAG). The IP_3 induces the release of Ca^{2+} from the endoplasmic reticulum [13, 88] and a further increase in concentration of intracellular Ca^{2+} ($[Ca^{2+}]_i$) results from an increased uptake of Ca^{2+} [66]. This flux in Ca^{2+} provides one of the intracellular signals required for activation. The formation of DAG together with Ca^{2+} leads to the activation and translocation of PKC. The generation of cAMP as well as increases in $[Ca^{2+}]_i$ and PKC all occur within the first 30 min after receptor occupation, thereby providing ample opportunity for cross-talk to occur.

This cross-talk can result in either an attenuation or amplification of the two diverse second messengers elicited. Indeed, there is evidence in both lymphocytes [5, 24, 63] and other cells [51, 52, 62, 91] that such interactions occur among second messengers. By altering the magnitude and kinetics of these second messengers, other biochemical and molecular perturbations are induced which dramatically alter the function of the lymphocyte. Increasing the concentration of cAMP in T cells also stimulated via the TCR generally results in a decrease in the proliferative capacity of these cells [61]. However, the interaction of neurohormones with lymphocytes does not always result in down-regulation of function. For example, prolactin can enhance in vitro and in vivo immune responses and lymphocyte function [11, 38]. This suggests that second messengers and subsequent biochemical events elicited by prolactin and a concurrent immune stimulus are facilitatory rather than antagonistic. It is interesting to note that there is evidence which indicates that binding of prolactin to its receptor on lymphocytes results in activation of PKC [79]. Precisely how similar second messengers that are generated by a neurohormone or immune stimulus binding to a lymphocyte interface with one another, thereby altering cellular function, remains to be determined.

In proposing a two-signal model of neural-immune interaction we chose a simplistic model which ignores many of the confounding features of both the neuroendocrine and immune systems. The constraints we place on this model, i.e. resting T cells receiving the two simultaneous signals, are unlikely to occur with a high probability under in vivo conditions. Such a static example obviously ignores the dynamics of both the endocrine and immune systems. Thus, the circulating levels of neurohormones vary and, concurrently, a given number of lymphocytes are constantly undergoing activation. The regulatory complexity of each system is enormous and becomes even greater when considering how they might interact. This complexity can be placed in the context of one- or two-signal models. Paramount to this is the fact that these signals can occur temporally out of phase. An antigen-stimulated T cell moving through the cell cycle may receive at any time a signal generated by neurohormone, assuming the activated T cell is expressing the receptor for the neurohormone. This interaction with neurohormone may alter the functional potential of the activated T cell. Such alteration can be either additive or synergistic with the magnitude governed by the strength of the signals. A similar scenario can be envisioned when the temporal sequence of signaling is reversed, i.e. neurohormone binding initially (signal one) followed by an immune stimulus (signal two). The limits placed on the potential for this temporal sequence to achieve a meaningful modification in lymphocyte function is determined by the ability of the lymphocyte to adapt to the biochemical and molecular events induced by the neurohormone binding to its receptor prior to antigen signaling. Generally, adaptation to a neurohormone signal occurs within minutes to a few hours. Examples of this form of a two-signal model are found for neurohormones. For example, preincubation of rat spleen cells with prolactin for 18 h followed by the addition of IL-2 induces proliferation in these cells over that seen with cells incubated with either prolactin or IL-2 [70]. Similarly, the production of superoxide anion by polymorphonuclear cells is greatly increased if these cells are preincubated for 3 h with growth hormone before stimulation with a phorbol ester [45]. These data serve to illustrate several other points concerning the proposed signaling models. Delivery of the first signal 3 h before the second signal is within the temporal limits required for these two signals to interact. However, preincubation for 18 h with a neurohormone, i.e. prolactin followed by the second signal is beyond the time that neurohormones are capable of modulating immune function. An alternative which is compatible with the proposed signaling models is that prolactin acts on a subset of preactivated T cells and renders them able to express the IL-2R. Thus, the lymphocytes have already received signal two as a result of an immune stimulus. It is important to recall that we previously discussed the fact that T cells require more than one signal to proceed through the cell cycle. Finally, these collective data demonstrate a synergistic response in the presence of the neurohormone and immune stimulus signals. Thus, the biologic responses elicited by the neurohormone and immune stimulus is significantly greater than the sum of responses produced by either the neurohormone or immune stimulus. This would strongly suggest that these two signals can cross-talk at the biochemical and molecular levels. Furthermore, different types of signal amplification can occur in one- or two-signal models [60]. One form of signal amplification, termed magnitude amplification, is characterized by a small concentration of stimulus eliciting a large number of effector molecules. For example, in a one-signal model a small concentration of ACTH generates a large number of cAMP molecules in the lymphocyte. The other form of signal amplification, sensitivity amplification, is characterized by a greater than expected increase in the response than anticipated from the percentage change in the stimulus. Sensitivity amplification appears to be operant in the two-signal model in which a synergistic effect on the lymphocyte function is measured. Thus, the binding of a neurohormone and an immune stimulus to their respective receptors on lymphocytes may result in a far greater change in second messengers than anticipated from a given concentration of either stimulus alone. To illustrate this, the binding of a certain concentration of ACTH to its receptor on a lymphocyte will increase the concentration of cAMP in the cell (one-signal model), yet this same concentration of ACTH will produce synergistically enhanced concentrations of cAMP when the second signal in the form of the immune stimulus is delivered (two-signal model). This form of signal amplification has enormous physiologic significance for neuroendocrine-immune interactions. It would suggest that large increases in the circulating levels of a neurohormone are not required to effect a change in a particular lymphocyte function. Only a small increase in neurohormone above normal background levels results in a synergistic increase in the second messenger elicited by the neurohormone binding to its receptor in the presence of immune signaling. We [24, 76, 77] have begun experiments to test this hypothesis using resting human T cells stimulated with isoproterenol to activate the β-adrenergic receptor (signal one) and PHA or anti-CD3 mab to provide the 'immune' stimulus (signal two). Stimulation of T cells with increasing concentrations of isoproterenol results in a progressive increase in cAMP with maximal values obtained at 10^{-5} *M* isoproterenol. PHA does not elicit cAMP to any appreciable level but rather activates PLC which results in the turnover of PI and generation of increased $[Ca^{2+}]_i$ and activation of PKC [1]. When T cells are exposed to both isoproterenol and PHA or anti-CD3 mab (a two-signal model) there is a synergistic rise in cAMP and a reduced proliferative response as compared to that observed with PHA stimulation alone [24]. These results suggest that a type of sensitivity amplification is taking place. This is further strengthened by the following unpublished observation: Stimulation of T cells with isoproterenol in concentrations ranging from 5×10^{-8} to 5×10^{-7} *M* does not result in significant cAMP production. T cells activated by PHA and stimulated with 5×10^{-8} *M* isoproterenol produce 10% of the maximum cAMP which results from co-stimulation with PHA and 10^{-5} *M* isoproterenol. A one-half log higher concentration (1×10^{-7} *M*) of isoproterenol yields 20% of maximum while 5×10^{-7} *M* isoproterenol plus PHA generates a cAMP response equal to that observed with 10^{-5} *M* isoproterenol and PHA. Thus, a one log increase (5×10^{-8} to 5×10^{-7} *M*) in the concentration of isoproterenol which singularly do not elicit a cAMP response, does induce the same amount of cAMP obtained with 10^{-5} *M* isoproterenol and PHA when co-stimulated with mitogen. These data would suggest that if sensitivity amplification is operating in a two-signal model, small increases in the concentration of neurally-derived substances above physiologic levels results in large alterations in lymphocyte function. Thus, this form of amplification has the potential to provide an enormous advantage to neural-immune interactions and correlates well with the role of these interactions in immune homeostasis. It will be interesting to determine if a similar phenomenon occurs with neurohormones and an immune stimulus.

Neurohormone Modulation of Immune Function – Future Directions

The evidence is overwhelming that neuropeptides can modulate immune function both in vivo and in vitro. What is not clear is how this is effected. In this section we will define some of the questions which are amenable to analysis and whose answers thereto will help establish the biological significance of this particular form of neural-immune regulation.

It would seem reasonable that if neurohormones are to interact effectively with and modulate lymphocyte or macrophage function, these cells should have high affinity receptors for the various peptides. As previously discussed, membrane receptors for a number of these peptides have been demonstrated on lymphocytes; however, confirmation of some of these receptors is still lacking. Additionally there is a paucity of information on the cellular distribution of these receptors. These data would greatly aid in deciding what types of functional assays should be performed. Moreover, it is equally important to gain an understanding of factors involved in the expression and modulation of these receptors. For example, is the density or affinity of these receptors altered when lymphocytes are activated? Another important missing piece of information is whether the molecular structure of the receptor present on lymphocytes is identical to that found on the target cells for the hormone. The advent of antibodies to these receptors will

Table 2. Chain of evidence for neuroendocrine-immune interactions

Motive	Immune homeostasis
Opportunity	Availability of neurohormones and presence of neurohormone receptors on cells of the immune system
Means	Modulates biochemical and molecular events involved in lymphocyte function

provide a valuable tool for determining the number of receptor-positive cells as well as the temporal expression of the receptors in relationship to lymphocyte activation.

With the rigorous establishment of membrane receptors for neurohormones on lymphoid cells, attention can then be focused on their mode of action. As previously discussed, there are ample data describing the phenomenologic events which occur when various neurohormones are added to cultures of lymphoid cells. What is lacking for the most part is an in-depth analysis of the biologic, biochemical and molecular events which transpire as a consequence of the hormone binding to its lymphocyte receptor. The use of one- and two-signal models and beyond may prove helpful in unraveling this complex process.

Conclusions

In this review we have presented a chain of evidence establishing motive, opportunity and means for interaction between the neuroendocrine and immune systems (table 2). The motive is to enable neurohormones to exert homeostatic control over immune function. This can take the form of either facilitatory or inhibitory modulations in immunity. Under normal physiologic conditions the magnitude and duration of this immunomodulation will be limited because the immune system has developed finely-tuned networks and circuits which provide for efficient control. Thus, neurohormones will not have a major influence on immune function but rather an associative and governing role. The opportunity for this form of neuroimmunomodulation to occur is provided by the availability of neurohormones to encounter cells of the immune system in concentrations sufficient to bind to functional receptors on these cells. Because lymphoid and myeloid cells reside in a number of different anatomical sites, their availability to neurohormones will differ. Finally, the means for neurohormonal modulation of immunity occurs as a consequence of neurohormones binding to their receptors on lymphoid and myeloid cells and subsequent induction of biochemical and molecular events which alter cellular function. We have discussed one- and two-signal models which can as first approximations be employed to explore these receptor-ligand interactions. It is important to stress that there is no conclusive evidence indicating that a neurohormone binding to its receptor on a resting lymphocyte (one-signal model) will induce this cell to proliferate or differentiate. This is consistent with the observations indicating that a lymphocyte must receive a number of immune-derived signals to achieve full function.

References

1 Altman, A.; Coggeshell, K.M.; Mustelin, T.: Molecular events mediating T cell activation. Adv. Immunol. *48:* 227–360 (1990).

2 Arrenbrecht, S.: Specific binding of growth hormone to thymocytes. Nature *242:* 255–257 (1974).

3 Baroni, C.D.; Fabris, N.; Bertoli, G.: Effects of hormones on development and function of lymphoid tissues. Synergistic action of thyroxin and somatotropic hormone in pituitary dwarf mice. Immunology *17:* 303–313 (1969).

4 Baroni, C.D.; Pesando, P.C.; Bertoli, G.: Effects of hormones on development and function of lymphoid tissue. II. Delayed development of immunological capacity in pituitary dwarf mice. Immunology *21:* 455–461 (1971).

5 Beckner, S.K.; Farrar, W.L.: Inhibition of adenylate cyclase by IL-2 in human T lymphocytes is mediated by protein kinase C. Biochem. Biophys. Res. Commun. *145:* 176–182 (1987).

6 Bellussi, G.; Muccioli, G.; Ghe, C.; DiCarlo, R.: Prolactin binding sites in human erythrocytes and lymphocytes. Life Sci. *41:* 951–959 (1987).

7 Berczi, I.: The effects of growth hormone and related hormones on the immune system; in Berczi (ed.): Pituitary Function and Immunity, pp. 134–159 (CRC Press, Boca Raton 1986).

8 Berczi, I.: Immunoregulation by pituitary hormones; in Berczi (ed.): Pituitary Function and Immunity, pp. 227–240 (CRC Press, Boca Raton 1986).

9 Berczi, I.; Nagy, E.: Prolactin and other lactogenic hormones; in Berczi (ed.): Pituitary Function and Immunity, pp. 162–183 (CRC Press, Boca Raton 1986).

10 Berkenbosch, F.; Oers, J. van; Rey, A. del; Tilders, F.; Besedovsky, H.: Corticotropin-releasing factor-producing neurons in the rat activated by interleukin-1. Science *238:* 524 (1987).

11 Bernton, E.W.: Prolactin and immune host defenses. Prog. Neuroendocrinimmunol. *2:* 21–29 (1989).

12 Bernton, E.W.; Beach, J.E.; Holaday, J.W.; Smallridge, R.C.; Fein, H.C.: Release of multiple hormones by a direct action of interleukin-1 on pituitary cells. Science *238:* 519 (1987).
13 Berridge, M.J.; Irvine, R.F.: Inositol triphosphate: a novel messenger in cellular signal transduction. Nature *312:* 315–321 (1984).
14 Besedovsky, H.; DelRay, A.; Sorkin, E.: Regulatory immune-neuro-endocrine feedback loops; in Berczi (ed.): Pituitary Function and Immunity, pp. 241–249 (CRC Press, Boca Raton 1986).
15 Besedovsky, H.; DelRay, A.; Sorkin, E.; Dinarello, C.A.: Immunoregulatory feedback between interleukin-1 and glucocorticoid hormones. Science *233:* 652–654 (1986).
16 Besedovsky, H.D.; Sorkin, E.: Immunologic-neuroendocrine circuits: physiological approaches; in Ader (ed.): Psychoneuroimmunology, pp. 545–574 (Academic Press, New York 1981).
17 Blalock, J.E.: The immune system as a sensory organ. J. Immunol. *132:* 1067–1070 (1984).
18 Blalock, J.E.: Production of neuroendocrine peptide hormones by the immune system. Prog. Allergy, vol. 43, pp. 1–13 (Karger, Basel 1988).
19 Blalock, J.E.; Smith, E.M.: A complete regulatory loop between the immune and neuroendocrine systems. Fed. Proc. *44:* 108–111 (1985).
20 Blalock, J.E.; Smith, E.M.; Meyer, W.J.: The pituitary-adrenocortical axis and the immune system. Clin. Endocrinol. Metab. *14:* 1021–1038 (1985).
21 Bost, K.L.: Hormone and neuropeptide receptors on mononuclear leukocytes. Prog. Allergy *43:* 68–83 (1988).
22 Brooks, W.H.; Cross, R.J.; Roszman, T.L.; Markesbery, W.R.: Neuroimmunomodulation: neural anatomical basis for impairment and facilitation. Ann. Neurol. *12:* 56–61 (1982).
23 Brown, S.L.; Smith, L.R.; Blalock, J.E.: Interleukin-1 and interleukin-2 enhance proopiomelanocortin gene expression in pituitary cells. J. Immunol. *139:* 3181 (1987).
24 Carlson, S.L.; Brooks, W.H.; Roszman, T.L.: Neurotransmitter-lymphocyte interactions: dual receptor modulation of lymphocyte proliferation and cAMP production. J. Neuroimmunol. *24:* 155–162 (1989).
25 Carr, D.J.J.; Blalock, J.E.: Neuropeptide hormones and receptors common to the immune and neuroendocrine systems: bidirectional pathway of intersystem communication; in Ader, Felten, Cohen (eds): Psychoneuroimmunology, pp. 573–588 (Academic Press, San Diego 1991).
26 Chabaud, O.; Lissitzky, S.: Thyrotropin-specific binding to human peripheral blood monocytes and polymorphonuclear leukocytes. Mol. Cell. Endocrinol. *7:* 79–87 (1977).
27 Clarke, B.L.; Bost, K.L.: Differential expression of functional adenocorticotropic hormone receptors by subpopulation of lymphocytes. J. Immunol. *143:* 464–469 (1989).
28 Clevenger, C.V.; Russell, D.H.; Appasamy, P.M.; Prystovosky, M.B.: Regulation of interleukin-2-driven T-lymphocyte proliferation by prolactin. Proc. Natl. Acad. Sci. USA *87:* 6460–6464 (1990).
29 Comsa, J.; Leonhardt, H.; Schwartz, J.A.: Influence of the thymus-corticotropin-growth hormone interaction on the rejection of skin allografts in the rat. Ann. N.Y. Acad. Sci. *249:* 387–401 (1975).

30 Coutelier, J.P.; Kehrl, J.H.; Bellur, S.S.; Kohn, L.D.; Nokins, A.L.; Prabhakar, B.S.: Binding and functional effects of thyroid-stimulating hormone on human immune cells. J. Clin. Immunol. *10:* 204–210 (1990).
31 Cross, R.J.; Brooks, W.H.; Roszman, T.L.: Modulation of T-suppressor cell activity by central nervous system catecholamine depletion. J. Neurosci. Res. *18:* 75–81 (1987).
32 Cross, R.J.; Brooks, W.H.; Roszman, T.L.; Markesbery, W.R.: Hypothalamic-immune interactions. Effects of hypophysectomy on neuroimmunomodulation. J. Neurol. Sci. *53:* 557–566 (1982).
33 Cross, R.J.; Campbell, J.C.; Markesbery, W.R.; Roszman, T.L.: Transplantation of syngeneic pituitary graft fails to restore T-cell function and thymus morphology in aged mice. Mech. Aging Dev. *56:* 11–22 (1990).
34 Cross, R.J.; Campbell, J.; Roszman, T.L.: Potentiation of antibody responsiveness after the transplantation of a syngeneic pituitary gland. J. Neuroimmunol. *25:* 29–35 (1989).
35 Cross, R.J.; Jackson, J.C.; Brooks, W.H.; Sparks, D.L.; Markesbery, W.R.; Roszman, T.L.: Neuroimmunomodulation: Impairment of humoral immune responsiveness by 6-hydroxydopamine treatment. Immunology *57:* 145–152 (1986).
36 Cross, R.J.; Markesbery, W.R.; Brooks, W.H.; Roszman, T.L.: Hypothalamic-immune interactions. I. The acute effect of anterior hypothalamic lesions on the immune response. Brain Res. *196:* 79–87 (1980).
37 Cross, R.J.; Markesbery, W.R.; Brooks, W.H.; Roszman, T.L.: Hypothalamic-immune interactions. Neuromodulation of natural killer activity by lesioning of the anterior hypothalamus. Immunology *51:* 399–405 (1984).
38 Cross, R.J.; Roszman, T.L.: Neuroendocrine modulation of immune function: the role of prolactin. Prog. Neuroendocrinimmunol. *2:* 17–20 (1989).
39 Dann, J.A.; Wachtel, S.S.; Rubin, A.L.: Possible involvement of the central nervous system in graft rejection. Transplantation *27:* 223–226 (1979).
40 Dumont, F.; Robert, F.; Bischoff, P.: T and B lymphocytes in pituitary dwarf Snell-Bagg mice. Immunology *38:* 23–31 (1979).
41 Duquesnoy, R.J.; Kalpaktsoglou, P.K.; Good, R.A.: Immunological studies on the Snell-Bagg pituitary dwarf mouse. Proc. Soc. Exp. Biol. Med. *133:* 201–206 (1970).
42 Enerback, L.; Lundin, P.M.; Mellgren, J.: Pituitary hormones elaborated during stress. Action on lymphoid tissues, serum proteins and antibody titres. Acta Pathol. Microbiol. Scand. *144:* suppl., pp. 141–144 (1961).
43 Fabris, N.; Pierpaoli, W.; Sorkin, E.: Hormones and the immunological capacity. III. The immunodeficiency disease of the hypopituitary Snell-Bagg dwarf mouse. Clin. Exp. Immunol. *9:* 209–225 (1971).
44 Fabris, N.; Pierpaoli, W.; Sorkin, E.: Hormones and the immunological capacity. IV. Restorative effects of developmental hormones of lymphocytes on the immunodeficiency syndrome of the dwarf mouse. Clin. Exp. Immunol. *9:* 227–240 (1971).
45 Fu, Y.K.; Arkins, S.; Wang, B.S.; Kelley, K.W.: A novel role of growth hormone and insulin-like growth factor-1. Priming neutrophils for superoxide anion secretion. J. Immunol. *146:* 1602–1608 (1991).
46 Gisler, R.H.; Schenkel-Hulliger, L.: Hormonal regulation of the immune response. II. Influence of pituitary and adrenal activity on immune responsiveness in vitro. Cell. Immunol. *2:* 646–657 (1971).

47 Goldowitz, D.; Scheff, S.; Cotman, C.: The specificity of reactive synaptogenesis: comparative study in the adult rat hippocampal formation. Brain Res. *170:* 427–441 (1979).

48 Hazum, E.; Chang, K.; Cuatrecassas, P.: Specific nonopiate receptors for beta-endorphin. Science *205:* 1033–1035 (1979).

49 Hemmick, L.M.; Bidlack, J.M.: Beta-endorphin modulation of mitogen-stimulated calcium uptake by rat thymocytes. Life Sci. *41:* 1971–1978 (1987).

50 Hemmick, L.M.; Bidlack, J.M.: Beta-endorphin stimulates rat T lymphocyte proliferation. J. Neuroimmunol. *29:* 239–248 (1990).

51 Ho, A.K.; Chik, C.L.; Klein, D.C.: Protein kinase C is involved in adrenergic stimulation of pineal cGMP accumulation. J. Biol. Chem. *262:* 10059–10064 (1987).

52 Ho, A.K.; Thomas, T.P.; Chik, C.L.; Anderson, W.B.; Klein, D.C.: Protein kinase C: subcellular redistribution by increased Ca^{2+} influx. J. Biol. Chem. *263:* 9292–9297 (1988).

53 Hough, C.J.; Halperin, J.I.; Mazorow, D.L.; Yeandle, S.L.; Millar, D.B.: Beta-endorphin modulates T-cell intracellular calcium flux and c-myc expression via a potassium channel. J. Neuroimmunol. *27:* 163–161 (1990).

54 Johnson, H.M.; Smith, E.M.; Torres, B.A.; Blalock, J.E.: Neuroendocrine hormone regulation of in vitro antibody production. Proc. Natl. Acad. Sci. USA *79:* 4171–4174 (1982).

55 Johnson, H.M.; Torres, B.A.: Immunoregulatory properties of neuroendocrine peptide hormones. Prog. Allergy, vol. 43, pp. 37–67 (Karger, Basel 1988).

56 Kalden, J.R.; Evans, M.M.; Irvin, W.J.: The effect of hypophysectomy on the immune response. Immunology *18:* 671–679 (1970).

57 Kavelaaro, A.; Eggen, B.J.L.; De Graan, P.N.E.; Gispen, W.H.; Heijnen, C.J.: The phosphorylation of the CD3 gamma chain of T lymphocytes is modulated by beta-endorphin. Eur. J. Immunol. *20:* 943–945 (1990).

58 Kelley, K.W.: Growth hormone, lymphocytes and macrophages. Biochem. Pharmacol. *38:* 705–713 (1989).

59 Kelley, K.W.; Brief, S.; Westly, H.J.; Novakofski, J.; Bechtel, P.J.; Simon, J.; Walker, E.B.: GH_3 pituitary adenoma cells can reverse thymic aging in rats. Proc. Natl. Acad. Sci. USA *83:* 5663–5667 (1986).

60 Koshland, D.E.; Goldbeter, A.; Stock, J.B.: Amplification and adaptation in regulatory and sensory systems. Science *217:* 220–225 (1982).

61 Krammer, G.: The adenylate cyclase-cAMP-protein kinase A pathway and regulation of the immune response. Immunol. Today *9:* 222–229 (1988).

62 Lamers, J.M.J.; Stinis, H.T.; de Jonge, H.R.: On the role of cyclic AMP and Ca^{2+}-calmodulin-dependent phosphorylation in the control of $(Ca^{2+}+Mg^{2+})$-ATPase of cardiac sarcolemma. FEBS Lett. *127:* 139–143 (1981).

63 Lerner, A.; Jacobson, B.; Miller, R.A.: Cyclic AMP concentrations modulate both calcium and hydrolysis of phosphatidylinositol phosphates in mouse T lymphocytes. J. Immunol. *140:* 936–940 (1988).

64 Lesniak, M.A.; Gorden, P.; Roth, J.; Gavin, J.R.: Binding of ^{125}I-human growth hormone to specific receptors in human cultured lymphocytes. J. Biol. Chem. *249:* 1661–1667 (1974).

65 Lesniak, M.A.; Roth, J.; Gorden, P.; Gavin, J.R.: Human growth hormone radioreceptor assay using cultured human lymphocytes. Nature *241:* 20–22 (1973).

66 Lichtman, A.H.; Segel, G.B.; Lichtman, M.A.: The role of calcium in lymphocyte proliferation (an interpretive review). Blood *61:* 413–422 (1983).
67 Lundin, P.M.: Action of hypophysectomy on antibody formation in the rat. Acta Pathol. Microbiol. Scand. *48:* 351–357 (1960).
68 MacLean, D.; Reichlin, S.: Neuroendocrinology and the immune process; in Ader (ed.): Psychoneuroimmunology, pp. 405–428 (Academic Press, New York 1981).
69 Mehriski, J.N.; Mills, I.H.: Opiate receptors on lymphocytes and platelets. Clin. Immunol. Immunopathol. *27:* 240–249 (1983).
70 Mukherjee, P.; Mastro, A.M.; Hymer, W.C.: Prolactin induction of interleukin-2 receptors on rat splenic lymphocytes. Endocrinology *126:* 88–94 (1990).
71 Nagy, E.; Berczi, I.: Immunodeficiency in hypophysectomized rats. Acta Endocrinol. *89:* 530–537 (1978).
72 Okouchi, E.: Thymus, peripheral tissue and immunological responsiveness of the pituitary dwarf mouse. J. Physiol. Soc. Japan *38:* 325–335 (1976).
73 Payan, D.G.; McGillis, J.P.; Goetzl, E.: Neuroimmunology. Adv. Immunol. *39:* 299–323 (1986).
74 Plaut, M.: Lymphocyte hormone receptors. Annu. Rev. Immunol. *5:* 621–669 (1987).
75 Prentice, E.D.; Lipscomb, H.; Metcalf, W.K.; Sharp, J.G.: Effects of hypophysectomy on DNCB-induced contact sensitivity in rats. Scand. J. Immunol. *5:* 955–961 (1976).
76 Roszman, T.L.; Carlson, S.L.: Neural-immune interactions: circuits and networks. Prog. Neuroendocrinimmunol. *4:* 69–78 (1991).
77 Roszman, T.L.; Carlson, S.L.: Neurotransmitters and molecular signalling in the immune response; in Ader, Felten, Cohen (eds): Psychoneuroimmunology, pp. 311–335 (Academic Press, San Diego 1991).
78 Roszman, T.L.; Cross, R.J.; Brooks, W.H.; Markesbery, W.R.: Neuroimmunomodulation: effects of neural lesions on cellular immunity; in Guillemin, Cohn, Melnechuk (eds): Neural Modulation of Immunity, pp. 95–109 (Raven Press, New York 1985).
79 Russell, D.H.: New aspects of prolactin and immunity: a lymphocyte-derived prolactin-like product and nuclear protein kinase C activation. Trends Pharmacol. Sci. *10:* 40–44 (1989).
80 Russell, D.H.; Kibler, R.; Matrisian, L.; Larson, D.; Poulos, B.; Magun, B.E.: Prolactin receptors on human T and B lymphocytes: antagonism of prolactin binding by cyclosporine. J. Immunol. *134:* 3027–3031 (1985).
81 Russell, D.H.; Matrisian, L.; Kibler, R.; Larson, D.F.; Poulos, B.; Magun, B.E.: Prolactin receptors on human lymphocytes and their modulation by cyclosporine. Biochem. Biophys. Res. Commun. *121:* 899–906 (1984).
82 Sapolsky, R.; Rivier, C.; Yamamoto, G.; Plotsky, P.; Vale, W.: Interleukin-1 stimulates the secretion of hypothalamic corticotropin-releasing factor. Science *238:* 522 (1987).
83 Saxena, Q.B.; Saxena, R.K.; Adler, W.H.: Regulation of natural killer activity in vivo. III. Effect of hypophysectomy and growth hormone treatment on the natural killer activity of the mouse spleen cell population. Int. Arch. Allergy Appl. Immunol. *67:* 169–174 (1982).
84 Schneider, G.B.: Immunological competence in Snell-Bagg pituitary dwarf mice: response to the contract-sensitizing agent oxazolone. Am. J. Anat. *145:* 371–380 (1976).

85 Sibinga, N.E.S.; Goldstein, A.: Opioid peptides and opioid receptors in cells of the immune system. Annu. Rev. Immunol. *6:* 219–249 (1988).
86 Smith, E.M.; Brosan, P.; Meyer, W.J.; Blalock, J.E.: A corticotropin (ACTH) receptor on human mononuclear lymphocytes: correlation with adrenal ACTH receptor activity. N. Engl. J. Med. *317:* 1266–1269 (1987).
87 Stein, M.; Schiavi, R.C.; Camerino, M.: Influence of brain and behavior on the immune system. Science *191:* 435–440 (1976).
88 Streb, H.; Irvine, R.F.; Berridge, M.J.; Schulz, I.: Release of Ca^{2+} from a nonmitochondrial intracellular store in pancreatic acinar cells by inositol-1,4,5-triphosphate. Nature *306:* 67–69 (1983).
89 Thrasher, S.G.; Bernardis, L.L.; Cohen, S.: The immune response in hypothalamic-lesioned and hypophysectomized rats. Int. Arch. Allergy Appl. Immunol. *41:* 813–820 (1971).
90 Tyrey, L.; Nalbandov, A.V.: Influence of anterior hypothalamic lesions on circulating antibody titres in the rat. Am. J. Physiol. *22:* 179–185 (1972).
91 Yoshimasa, T.; Sibley, D.J.; Bouvier, M.; Lefkowitz, R.J.; Caron, M.G.: Cross-talk between cellular signaling suggested by phorbol-ester-induced adenylate cyclase phosphorylation. Nature *327:* 67–70 (1987).

Thomas L. Roszman, Department of Microbiology and Immunology,
University of Kentucky Medical Center, Lexington, KY 40536 (USA)

Subject Index